D0758883

MALFORMED FROGS

MALFORMED
FROGS

The Collapse of Aquatic Ecosystems

Michael Lannoo

UNIVERSITY OF CALIFORNIA PRESS   Berkeley   Los Angeles   London

University of California Press, one of the most
distinguished university presses in the United States,
enriches lives around the world by advancing
scholarship in the humanities, social sciences, and
natural sciences. Its activities are supported by the
UC Press Foundation and by philanthropic
contributions from individuals and institutions.
For more information, visit www.ucpress.edu.

University of California Press
Berkeley and Los Angeles, California

University of California Press, Ltd.
London, England

©2008 by
The Regents of the University of California

Library of Congress Cataloging-in-Publication Data

Lannoo, Michael
    Malformed frogs / Michael Lannoo.
        p.   cm.
    Includes bibliographical references.
    ISBN 978-0-520-25588-3 (cloth : alk. paper)
    1. Frogs—Abnormalities. 2. Frogs—Research.
I. Title.
    QL668.E2L298 2008
    571.9'51789—dc22

                                    2008004005

Manufactured in the United States of America
16   15   14   13   12   11   10   09   08
10   9   8   7   6   5   4   3   2   1

The paper used in this publication meets the
minimum requirements of ANSI/NISO
Z39.48-1992 (R 1997) (Permanence of Paper). ∞

Cover illustration: Two specimens of *Rana pipiens*
with multiplied limb elements. Specimens collected
by the Minnesota Pollution Control Agency.
Photos by the author.

QL
668
.E2
L298
2008

Dedicated to the memory of Dan Sutherland

Dan after sampling the ROI site in central Minnesota, August 2001.
Photo by the author.

Royalties from this book will be donated to the Dan Sutherland Memorial
Scholarship administered by the University of Wisconsin, La Crosse Foundation.

And I brought ye into a plentiful country, to eat the fruit thereof and the goodness thereof; but when ye entered, ye defiled my land, and made mine heritage an abomination.

JEREMIAH 2:7

We humans may be the smartest objects that ever came down the pike of life's history on earth, but we remain outstandingly inept in certain issues, particularly when our emotional arrogance joins forces with our intellectual ignorance.

STEPHEN JAY GOULD
Lying Stones of Marrakech, 2000

CONTENTS

PREFACE

Attending scientific meetings may be the least professional thing that scientists do. Meetings are a lot about personal interactions, and the outcomes of personal interactions are dependent on attributes such as gender (males tend to have an advantage), size (increased height is more influential than increased weight), volume of speech (the louder the better), quantity of speech (the more the better, to a point), and speed of speech (faster being exponentially better). A quick consideration of the scientific method will instantly assure you that these qualities have nothing at all to do with scientific excellence. They do, however, have everything to do with the perception of scientific excellence. Further, people are more comfortable speculating about issues and giving personal opinions when talking at meetings than they are when writing papers for publication, which then become part of the permanent record (even if ignored or forgotten). Therefore, much of what happens at scientific meetings is non-science, although there can be no doubt that some non-scientific activities enable scientific progress.

I have disappointingly few personal attributes that give me social advantage, so at meetings I tend to sit in the back row and watch events, both scientific and non-scientific, play out in front of me; after all, I have been trained to observe. I also hang out with my friends and I try to meet people whom I do not know but whose work I have come to admire. I like meeting students—the difference between potential and eventual professional status is fascinating (because it is so often uncoupled). At meetings and elsewhere, I also carry with me some advice that I overheard my grandfather, who was an alderman, give my father. My grandfather would come home from city council meetings and exclaim, with a shot of Three Feathers in his left hand and a can of Hamm's in his right: "Donald, when you're dealing with people, always look for the game behind the play." A strange phrasing. The "game behind the play" didn't mean too much to me when I was in the third grade, but he said it so often and so forcefully in his immigrant Flemish/English that I carried it with me until I was finally old enough to understand; and now, every once in a while, I find that I need to remember it. By the way, although my grandfather was in no social or intellectual position to be aware of this, searching for the game behind the play is the same approach, flipped on its belly, taken by artists, composers, poets especially, and others with great creative minds.[1]

The first malformed frog meeting held in the United States, called "The Workshop on Central North American Amphibian Deformities," was sponsored by the Environmental Protection Agency's Mid-Continent Ecology Lab in Duluth in the fall of 1996. Rather than watching from my usual spot in the cheap seats, I found myself front and center as I led one of the two breakout groups during the afternoon of the first day. Focused meetings, especially those organized to address hot-button issues such as environmental train wrecks, make scientific meetings even more non-scientific. When you add (1) agendas (both personal and agency driven); (2) tension (people vs. people, federal vs. state, federal vs. federal, government vs. academics, government vs. NGOs, NGOs vs. academics, academics vs. academics, and so on and so on); (3) the presence of "ambulance chasers" (in scientific circles these are folks who have a knack for writing successful grant proposals and shifting their research emphasis on a dime, depending on funding opportunities); and (4) the potential to obtain large amounts of grant money, to all of the personal attribute issues, you almost always have a big, confusing mess. Imagine the chaos of people leaving a burning theater, in reverse.

As the co-chair of the U.S. Working Group of the Amphibian Specialist Group, created under the guidance of the World Conservation Union's (IUCN) Species Survival Commission (SSC), I have (as much as I can help it) no personal or professional agenda except to see that we solve amphibian decline and malformation problems using the best possible science.[2] I have seen some first-rate scientists fail because the same personal attributes that created their success—mammoth egos and immense self-confidence—

sealed their fate as they fell too much in love with themselves and their ideas, and could not let go. This collapse always involves a lack of credibility, which vanishes—sometimes gradually, sometimes all at once—as these once-fine scientists create a Rube Goldberg construct of logic (or if you prefer, a tangled web) to explain contrary facts, in what always turns out to be a failed salvage attempt of their pet theory. Time is the only variable in this scientific inevitability. Careers are like oak trees; it takes a long time, good, hard work, and maybe a little luck, to grow an impressive one, and a single afternoon of shortsightedness to cut it down.

In Duluth, after that first day of discussing malformed frogs, and while waiting for friends at the bar in a trendy waterfront restaurant, I realized that I was stunned. Sitting alone and drinking a better beer than my grandfather had ever considered, I promised myself that I would wait at least ten years before weighing in on the malformed frog problem. In the meantime, I would see as many malformed frogs as I could, help wherever I could, and keep paying attention. But I would not play for keeps until the field thinned and the issues cleared; after I had had a chance to assess the data, form my own impressions, and weigh them against what everyone else was saying. And, to my great surprise (because, just like everybody else's, none of my plans ever seem to follow their script), that is just about what happened. I only emerged from my personally imposed storm cellar (where we in the midwest go to hunker down) to formally work on malformed frogs three times. First, in 1998 I chaired the Malformity and Disease session held at the Midwest Declining Amphibian Conference in Milwaukee.[3] Second, Bill Souder interviewed me, often at length, for his book "A Plague of Frogs."[4] And finally, in 2001, Dan Sutherland and I undertook a two-week "expedition" (a much too formal term to describe what we did) to resample the hottest of the Minnesota hotspots,[5] a memory with an old friend that I will always cherish.

Now, almost exactly 10 years after Duluth, and having examined hundreds of malformed frogs for the Minnesota Pollution Control Agency, hundreds of frogs for the U.S. Fish and Wildlife Service's multi-year survey of malformed frogs on our National Wildlife Refuges, and hundreds of malformed frogs for dozens of friends, I will open the storm cellar doors for good, haul out my "x-rays," turn on the light box, and repeat Rafiki's words to Simba:[6] "It is time."

ACKNOWLEDGMENTS

This book wrote itself in a matter of a few months. Given the often demonstrated inverse relationship between time spent and errors generated, and the opposite, direct relationship between eyes observing and perspective achieved (faculty meetings excepted), I owe a huge debt of gratitude to Richard Wassersug, Alisa Gallant, Dave Hoppe, Bill Souder, Joe Mitchell, Tony Mescher, Bob McKinnell, Priya Nanjappa-Mitchell, Alan Resetar, Rob Lovich, Joe Keisecker, Ken Lang, Bill Cummings, Brian Todd, Peter Johnson Lannoo, and Susan Johnson Lannoo for offering suggestions on early drafts. I thank Danette Pratt for creating the graphics and standardizing the presentation of the radiographic images, and Donna Helfst for copy editing the ms. Thanks to Bob Klaver, who assembled several of the digital maps, to Dan Fogell and Dave Hoppe for permission to use photographs, and to Scott Norton for turning this manuscript into a book. Thanks also to all of the researchers and curators who trusted their specimens to my care—I would not have these data without you. I had insightful conversations with Carol Meteyer, David E. Green, David M. Green, Gary Casper, Tony Neff, Louise Rollins-Smith, Dave Bradford, John Crawford, Bill Peterman, Phil Tevis, and Pastor Mike Rozumalski at West Denmark's Folk School about various aspects of this project. Finally, I thank Joe Eastman, who kindly opened up his lab to me and my coolers full of malformed frogs, kept the coffee pot filled, and manned the dark room. As always, when thinking about my closest colleagues, I'm reminded of the words of John Steinbeck and Ed Ricketts: "We sat on a crate of oranges and thought what good [people] most biologists are . . ."[1]

Funding for this work was generously provided by the Minnesota Pollution Control Agency, the U.S. Fish and Wildlife Service, Indiana University School of Medicine, and NIH contract grant #NS376000-01.

INTRODUCTION

It is always easier to identify problems than
to construct solutions.

As a cultural phenomenon it began in the summer of 1995 in south-central
Minnesota, with school kids on a field trip.[2] While exploring a rural wetland, Cindy Reinitz and her junior high school students discovered a large
number of northern leopard frogs (*Rana pipiens*) that were having trouble jumping. When normal frogs with two good hind legs jump they
maintain control in the air and land on their hands and their chest. But
the leopard frogs these kids were finding had missing legs or missing parts
of legs, and when leopard frogs with only one good leg jump, they oftentimes lose control, twist, and land on their backs, exposing the whites of
their bellies. Walk along the edge of a wetland with jumping malformed
leopard frogs and it appears to your brain as if you've gotten punched in
the nose—you see a lot of flashes of white.

Next to the morning sickness drug thalidomide,[3] and perhaps Hooker
Chemical Company's dump site at Love Canal[4] (both of which affected
humans), no environmental tragedy causing grotesque bodies or body parts
has grabbed society's preoccupation with itself and shaken it quite like the
malformed frog phenomenon. Malformed frogs penetrated the American
consciousness and perhaps, ever so slightly, its conscience. As expected,
there were political ramifications. Between 1998 and 2000, Secretary of
the Interior Bruce Babbitt created the U.S. Geological Survey's
Amphibian Research and Monitoring Initiative (ARMI),[5] not so much
(as is generally thought) in response to the larger problem of amphibian

declines as to the widespread presence of frog malformations and the fear they were causing among society at large.

A society in comfort can ignore, even dismiss, science as a tool for creating health and well being. But a society in crisis, frightened and facing illness, not only looks to science for answers, but *demands* answers from science (this behavior of societies is in striking contrast to the behavior of individuals, who often seek comfort in organized religion). The discovery by Cindy Reinitz' students was quickly followed by scores of other, similar reports. The malformed frog problem reached a tipping point and became a phenomenon. Society panicked—the State of Minnesota began distributing bottled water to families perceived to be in harm's way—and turned to science for answers.

In his book, "Triangle: The Fire that Changed America," David von Drehle points out (p. 197):[6] "A century ago, the idea of tackling social problems by collecting *facts*—as opposed to scriptural passages or philosophical tenets—was groundbreaking." The wilderness advocate and spiritualist Sigurd Olson voiced this same sentiment when he echoed Aldo Leopold in noting (p. 250):[7] "Science has made one contribution to culture which seems to me permanent and good: the critical approach to questions of fact." Indeed, no single idea has changed humanity, or the face of the Earth for that matter, more than the notion of a scientific method. The scientific method applied to medicine[8] created not only the Johns Hopkins School of Medicine, but also the model of Johns Hopkins that was, in the early portion of the twentieth century, applied to all accredited U.S. medical schools. This model still holds today.[9] Modern, scientific-based medicine applied throughout the world has lowered infant deaths, raised life expectancy (mostly because fewer kids are dying, not because healthy people are living longer) and created the human population boom that more than anything else drove both the excitement and the apprehension of the twentieth century.

The scientific method at its most basic (and as I learned it in sixth grade, when teaching science in America meant teaching science) consists of: (1) observation; (2) hypothesis formation, and (3) hypothesis testing. A simple concept, but it contains some nuances. For example, in a society ostensibly based on faith, the notion that facts trump ideas can be startling; so much so that we often need reminders to keep our objectivity on course. The famed teacher and author, Norman Maclean writes

(p. 137):[10] ". . . science start[s] and end[s] in observation, and theory should always be endangered by it." The agnostic and antagonistic Thomas Henry Huxley (also known as "Darwin's bulldog") said the same thing more succinctly when referring to Herbert Spencer's notion of Social Darwinism (p. 112):[11] "A beautiful theory, killed by a nasty, ugly little fact." Huxley repeated himself more eloquently in a letter to Charles Kingsley (p. 149):[12] "Sit down before fact as a little child, be prepared to give up every preconceived notion, follow humbly wherever and to whatever abyss nature leads, or you shall learn nothing." The motto of the world's oldest scientific society, the Royal Society (Huxley was a member) is *nullius in verba*—"facts, not assertions, are what matter."[13] Even Robert Frost, the United States' former poet laureate, observed:[14] "The fact is the sweetest dream that labor knows," demonstrating a broad understanding of the narrow gap between science and humanity.

A second nuance peculiar to the scientific method, one championed by the philosopher Karl Popper,[15] is that scientific theories can never directly be proven correct. Instead, theories become generally accepted as alternative theories are proven wrong. Stephen Jay Gould points out (p. 155):[11] "The only universal attribute of scientific statements resides in their potential fallibility. If a claim cannot be disproven, it does not belong to the enterprise of science." Norman Maclean, who it must be said understood science better than most scientists (but probably not better than Robert Frost), dismissed the notion that the deadly Mann Gulch wildfire blew up because the hot flames were explosively spread by cold downdrafts created by a stalled thunderhead as a sort of (p. 126):[10] ". . . all purpose theory you can't disprove." And as a litmus test of whether or not a tale belongs to the realm of science, Gould posits it does when it can (p. 44):[12] ". . . be tested with definite evidence against reasonable alternatives."

A third, nominal view of the scientific method is objectivity, something that does not come easily to humans. But this too, is nuanced. Gould notes, correctly I think (p. 104):[16]

. . . the mind's curiosity cannot be suppressed (Why would anyone ever want to approach a problem without this best and most distinctive tool of human uniqueness?). Therefore [as a scientist], you will have suspicions and preferences whether you acknowledge them or not. If you truly believe that you are making utterly objective observations, then you will easily tumble into trouble, for you will probably not recognize your own inevitable prejudices. But if you acknowledge a context by posing explicit questions to test (and, yes, by inevitably rooting for a favored outcome), then you will be able to specify—and diligently seek, however much you may hope to fail—the observations that can refute your preferences. Objectivity cannot be equated with mental blankness; rather, objectivity resides in recognizing your preferences and then subjecting them to especially harsh scrutiny—and also in a willingness to revise or abandon your theories when the tests fail (as they usually do).

The problem of course, for scientists and non-scientists alike, is admitting it when we are wrong. Again, Gould (p. 117):[11] "Surely, in science, it is no sin to be wrong for good reasons." He continues (p. 90):[12] "No one can possibly be right about everything the first time. When controversies pit first-rate scientists against one another, any resolution will take bits and pieces of all views. Who could be so misguided as to get everything absolutely wrong?"

This topic of objectivity and the scientific method is deeply fertile ground for commentary. Rather than considering it at length here, I offer three quips that encompass my feelings on the subject. First, in his book "Deep Survival," Laurence Gonzales observes (p. 258–259):[17] "Integrity begins with humility." Second, the visionary ecologist Ed Ricketts, in commenting on a paper by W. K. Fisher, wrote that it was (p. 272):[18] ". . . the work of a man who has too much integrity to make things fit, for convenience, that *won't* fit in fact." Third, I note Maclean's take on the portrait of the eminent fire scientist Charles Gisborne (p. 139):[10] "He looks you square in the eye but is half amused as if he had caught you too attached to one of your theories, or to one of his."

It has now been over a decade since those Minnesota school kids went on their infamous field trip and it would be both a fair assessment to say, and a surprise for the public to learn, that science has failed to solve the malformed frog phenomenon. This is true despite spending millions of public and private dollars and publishing many (but not as many as you might think) research papers, and in the face of careers being jumpstarted and careers being crippled. Why has science failed? Because the scientists forgot, or perhaps never knew, that science (p. 79):[12] ". . . dedicates itself above all to fruitful doing, not clever thinking; to claims that can be tested by actual research, not to exciting thoughts that inspire no activity." The goal of the malformed frog inquiry should have been to fix the problem. Because the problem has not been fixed, it is accurate to say that up to this point in time science has failed; I know of no single malformed frog hotspot that has been restored to ecological health due to either the direct or indirect actions of scientists.

The purpose of this book is to explore the science underlying the malformed frog phenomenon: its history, the background that science provides to our current thinking, and what science offers in terms of solutions to the problem. I begin, in Chapter 1 (What Is an Amphibian Malformation?), by defining amphibian malformations, describing the mechanisms underlying malformations, and outlining their developmental origins. In Chapter 2 (Malformed Frog Types), I present radiographs demonstrating the extensive range of frog malformation types. Frog malformations are much more than multiple hindlimbs, and to appreciate the broader problem we must fully appreciate the extent of malformation types. In Chapter 3 (Hotspots), I describe several of the hottest of the Minnesota malformed frog hotspots, as well as several other important sites. Some of these sites resemble the pond next door, others are truly skanky. Known causes of malformations are listed in Chapter 4 (Causes). To appreciate the importance of these causes, in Chapter 5 (Resolutions) I critically evaluate the data underpinning this knowledge. To reduce my own bias, I rely heavily on the raw data and the original wording presented in the featured papers. In Chapter 6 (Human Malformations and Causes), I present types and known causes of human congenital anomalies. It is important to remember that the reason malformed frogs generated so much public interest was the possibility that what was happening to frogs might be happening to humans. In the Appendix (Species Affected), I present a list of the 293 currently recognized North American amphibians and highlight the 71 species known to have exhibited malformations.

The net effect of the first six chapters and the appendix is to break down the data that bolster our scientific understanding of amphibian malformations, and to evaluate their overall significance. In Chapter 7 (Solutions), I re-assemble these data to offer a new construct. Some surprising perspectives are revealed. First, every seriously considered cause of amphibian malformations has flaws as a universal theory, due either to a lack of evidence or to contrary facts. Second, in the context of human congenital anomalies, the most important and overlooked fact in all of the malformed frog literature is that a subset of malformed frogs have chromosomal damage. This is a critical observation and suggests that in some cases, factors causing amphibian malformations might also be affecting humans. Third, for several reasons, some valid, some not, the priority of the malformed frog investigation has been to identify and promote causes of malformations. As I indicate above, this approach has been misguided. The focus of the malformed frog investigation should have been to solve the problem. Causes of malformations, while interesting, must take a back seat to remediation. And the key component of any remediation effort should be to control chemical *and* nutrient runoff into affected wetlands. The not-so-funny little joke behind the malformed frog investigation is that while much of the money and effort expended to date have focused on causes, remediation efforts will largely be the same no matter what the cause, and are therefore mostly independent of cause. Put another way, you do not need to know much about causes to solve the malformed frog problem.

This observation is not the same as concluding that solving the malformed frog problem will be easy. We do not know how easy or difficult solutions will be, because to date nobody has tried.

It is the job of science to lead the way in showing how to solve the malformed frog problem; the solutions themselves, however, must come from a willing society. When so many Americans seek moral guidance on environmental issues from Genesis 1:26 ("And God said 'Let *us* make man in our image, after *our* likeness: and let them have dominion over the fish of the sea, and over the fowl of the air, and over the cattle, and over all the earth, and over every creeping thing that creepeth upon the earth'"), perhaps a world populated with malformed frogs is about all we should expect. Maybe not. Look at a kid with a tiny arm and nubbins for fingers watching a little league baseball game, and the empty question "Can we do better?" becomes the determined statement "We must do better." As we will see, far from simply being a scientific failure, the malformed frog phenomenon represents a societal meltdown. And we should heed the warning, for as all of us intuitively know (whether our view of religion extends beyond the conventional to include kids and baseball, or not)— nature bats last.

ONE

WHAT IS AN AMPHIBIAN MALFORMATION?

> Symmetry is an unmistakable sign that there's relevant
> information in a place. That's because symmetry is a
> property shared by a relatively small number of things in
> the landscape, all of them of keen interest to us. . . .
> Symmetry is also a sign of health in a creature, since
> mutations and environmental stresses can easily disturb it.
>
> MICHAEL POLLAN[1]

The word "malformation" literally means "bad form." Bad form in most
animals means an unintended lack of symmetry, or an imbalance in structure, color, or other quality. A lack of symmetry can arise through one
of three mechanisms:[2]

Genetic   Genes are flawed, or the expression of genes during development is flawed. Albinism (in which animals have white bodies and pink
eyes) is a familiar genetically determined malformation. Albinism is often
caused by a mutation that results in the failure of embryonic neural crest cells
to migrate. Similar genetic mutations in northern leopard frogs produce
blue or partially blue axanthic individuals (Color Plate 1),[3] the kandiyohi
pigment morph (which has tight blotches rather than spots; Color Plate 2),
and the burnsi pigment morph (which lacks dorsal body spots; Color Plate
3).[4] These mutations are not fatal and in the cases of the kandiyohi and
burnsi morphs are caused by an autosomal dominant allele and need only
be passed on by one parent. Roughly 50% of all offspring between one of
these morphs and a normally pigmented leopard frog inherit the pattern;
from 75–100% of offspring between two of these pigment morphs inherit
the pattern. Pieter Johnson and his colleagues[5] have suggested using the
term "deformity" for malformations such as these that arise from genetic
mechanisms. Not everyone has followed this suggestion, although such
precision in thought guides understanding.

PLATE 2.   A green kandyohi northern leopard frog morph from western
Minnesota. Photo by Michael Redmer©/www.mikeredmer.com.

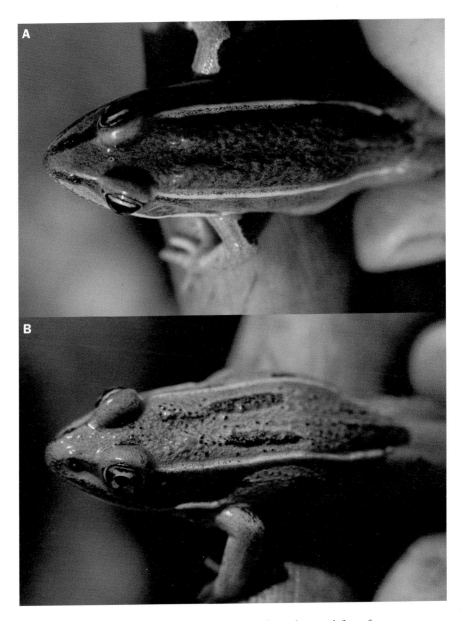

PLATE 3. (A) A brown burnsi morph northern leopard frog from northwestern Iowa. Note the lack of dorsal spotting. Northern leopard frogs may be either brown or green, and in the Upper Midwest, brown and green animals occur in about a 50:50 ratio. (B) A green burnsi morph northern leopard frog (or is it?—note the two elongate dorsal spots) from northwestern Iowa. This animal teaches a lesson: when working with amphibians we must be willing to consider variants that defy convenient characterization. Photos by the author.

PLATE 4. *Rana pipiens*. 50 mm SUL. Collected on 24 September 1997 at the SUN site in Ottertail County by Minnesota Pollution Control Agency field biologists. Note on the left thigh, proximal to the site where the limb segments are missing, the normal barred pattern along the dorsal hind limb (see right side of animal) is disrupted. Spotting is smaller and oriented along the long axis of the limb. Failed predation cannot produce this pigment pattern. Animals from this site are also shown in Figures 2.16 and 2.54.

other malformations. Hoppe's conclusion was unavoidable—these malformations were not genetically based, the cause was in the water.

In deciding between epigenetic and trauma-based causes, trauma oftentimes seems the least likely. Trauma can and does happen, and perhaps most trauma in amphibian larvae comes from failed predation attempts. However, we expect trauma to produce scarring, and most malformed animals show no signs of scarring or any other type of wound-healing process. This is true despite the fact that limb development, and therefore the opportunity for limb trauma, immediately precedes metamorphosis (remember that most malformed frogs are collected at or just after metamorphosis, when scarring should be obvious). Further, many frogs have malformations that simply cannot be due to trauma (e.g., eyes not in their proper location, multiple bent bones, multiple fluid-filled sacs, bloating, abnormal pigment patterns, small heads), or if they were due to trauma the injuries sustained (missing pelvic and spinal components) would likely be inconsistent with life. Forelimb malformations in newly metamorphosed animals are also unlikely to arise from trauma. For most of a tadpole's life forelimbs are tucked under enclosed gill coverings and therefore hidden from visually oriented predators, who are unlikely to selectively eat what they cannot see.

Given this evidence, the amphibian malformations that most concern us probably arise from epigenetic mechanisms—consequences of the environment in which genetic expression is occurring. As Michael Pollan has observed:[1] ". . . the ecological effects of changes to the environment often show up where we least expect to find them (p. 211)."

MOST AMPHIBIAN MALFORMATIONS ARE FROG MALFORMATIONS

Martin Ouellet assembled a comprehensive review of the literature on amphibian malformations[9] and from his study we can conclude that most amphibian malformations are frog and toad malformations. Ouellet's search found data on malformations occurring in more frog and toad species, sites, and specimens (Table 1.1).

The data my colleagues and I have collected from the United States support this conclusion. In the past decade we have radiographed 2,377 malformed, or suspected-to-be malformed, amphibians; 2,363 were frogs, 14 were salamanders. There were, however, sampling biases. In field protocols

TABLE 1.1.   *Summary of Martin Ouellet's data showing most amphibian malformations are frog and toad malformations*

	Frog and Toad vs. Salamander Numbers	Ratio
Species affected	67:26	2.6:1
Sites where found	159:43	3.7:1
Numbers collected	11,687:2,512	4.5:1

followed by both the Minnesota Pollution Control Agency (~ 702 specimens) and U.S. Fish and Wildlife Service (666 specimens through the 2005 field season), there was a stated emphasis on collecting malformed frogs rather than malformed salamanders. Therefore the species and sites sampled, the timing of sampling, and other practices favored the collection of frogs.

FROG SKELETONS

Because the form, or basic size and shape, of vertebrates is generated from their skeletons, and because skeletal tissue is the first to form in developing limbs, any serious description of malformations must include an examination of skeletons. There are several ways to examine skeletons, but we rely on radiographic (or roentgenographic in the old eponymous and cumbersome terminology[10]) analyses—we take "x-rays" (although, strictly speaking, x-rays are the type of radiation that produces the image on film; the actual image itself is termed a "radiograph"). To know what is abnormal, we must first know what is normal.

Frog skeletons are simple but variable (there are, after all, over 5,200 recognized species of frogs; in fact, the late Dr. Alan Holman used the shape of the hip joint [acetabulum; see below] to identify fossilized frog bones).[11] Figures 1.1 and 1.2 show labeled illustrations of frog and toad skeletons.[12, 13, 14] From these images, and following the recent review of Handrigan and Wassersug[15] we can see that frogs exhibit:

a large skull with a broad jaw and prominent eye sockets;
a rigid, shortened vertebral column with five to eight vertebrae (humans have 25);

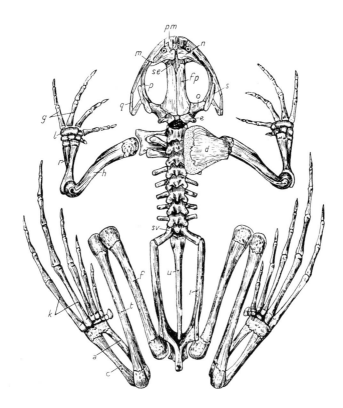

FIGURE I.I Frog skeleton (presumably *Rana* sp.). Left suprascapular and scapular removed. From J. Z. Young.[13] Letters code to bones as follows: *a*, tibiale (astralagus); *c*, fibulare (calcaneum); *d*, suprascapular; *e*, exoccipital; *f*, femur; *fp*, frontoparietal; *g*, metacarpals; *h*, humerus; *i*, ilium; *k*, metatarsals; *l*, carpus; *m*, maxilla; *n*, nasal; *o*, prootic; *p*, pterygoid; *pm*, premaxilla; *q*, quadratojugal; *r*, radioulna; *s*, squamosal; *se*, sphenethmoid; *t*, tibiofibula; *u*, urostyle; *sv*, sacral vertebra. Used with permission of Oxford Press.

> no ribs (this generalization holds for most frogs, including all species found in North America)
>
> relatively simple shoulder and pelvic girdles; and
>
> strong but simple limbs.

Shoulder girdles consist of sternal (breast) bones along the midline of the belly, scapular bones (shoulder blades) along the back, and coracoids and

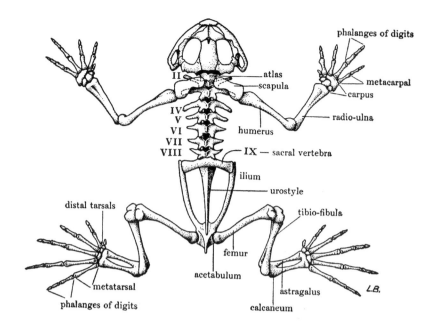

FIGURE 1.2   Skeleton of "common" toad (presumably *Bufo* sp.). From A.S. Romer.[14] In his caption, Romer notes: "Except in minor details, this is identical with the frog skeleton. (After Jammes.)" Used with permission of The University of Chicago Press.

clavicles (collar bones) connecting the sternal bones to the scapulars. The coracoid and scapular bones form the socket for the forelimb humerus.

Pelvic girdles consist of sacral, coccygeal, ischial, and pelvic bones aligned with the backbone, and paired iliums to either side (laterally). The acetabulum, or articulation between the pelvis and the femur of the leg, occurs at the junction of the ilium, ischium, and pelvic bones.

In frogs that jump, such as northern leopard frogs (see Fig. 1.1 and compare to Fig. 1.2), the pelvis and hindlimbs are modified to generate thrust; the shoulders and forelimbs are modified for landing. Male frogs may have larger forelimbs than females, to better grasp females during mating (amplexus).

Frog limbs are similar to the limbs of all vertebrates, including humans, in having the closest segment to the body supported by a single long bone (humerus in the arm, femur in the leg), and the next closest segment supported by two long bones (radius and ulna in the arm, tibia and fibula in the leg), which in frogs are fused to form the radioulna and tibiofibula, respectively. Wrist (carpal) and ankle (tarsal) bones are numerous but may be reduced or fused, depending on the species. Ankle bones include the elongated tibiale and fibulare. Hands are composed of metacarpal bones (the palm) and phalanges (fingers). Feet are composed of metatarsal bones (the foot in a restricted sense) and phalanges (toes). Four digits are expected on the hands, five digits are expected on the feet.

THE NATURE OF BONE

When we think of skeletons we think of bones. Bones, particularly long bones, consist of two parts: an outer, hard compact surface (periosteal bone), and an inner, chambered, cancellous core (marrow cavity).[16] Bones are vascularized (have a blood supply). Bone tissue consists of calcium supported by a protein matrix called hydroxyapatite. With life experiences, bones are remodeled and reshaped. This complicated process is accomplished by differential addition to or subtraction from the bone surface. Bone addition is achieved through the actions of cells called osteoblasts; bone subtraction is accomplished by osteoclasts.

Most bones, including the bones of the spine, shoulder and pelvic girdles, and limbs, develop from cartilage. Cartilage is unvascularized—it has no capillary network of its own—and is made of protein (collagen and proteoglycan complexes) and water. During the metamorphosis from an aquatic tadpole to a terrestrial juvenile, most portions of frog skeletons undergo a conversion from cartilage to bone. This conversion is triggered in part by the growth of arteries into the cartilaginous skeletal elements. Cartilage is much less dense than bone, so radiographs of frog tadpoles do not show the contrast between skeletal and non-skeletal elements seen in metamorphosed frogs (Figure 1.3).

HOW DO MALFORMATIONS ARISE
DURING DEVELOPMENT?

All vertebrate tissues develop from one of three embryonic germ layers: ectoderm (giving rise to epidermis, pigment cells, and nervous tissue), mesoderm (which creates the dermis of the skin, skeleton and muscle tissue), and

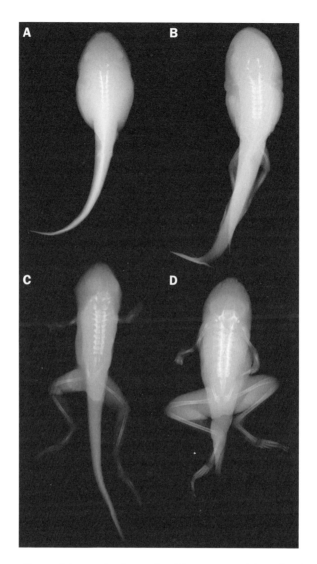

FIGURE 1.3   *Rana pipiens* tadpoles and mid-metamorphic animals collected in Minnesota during the 1996 and 1997 field seasons, illustrating
the partial sequence of bone ossification: (A) collected August 1997,
~39 mm SUL; (B) collected June 26 1996, ~31 mm SUL; (C) collected
August 1997, ~31 mm SUL; (D) collected June 26 1996, ~33 mm SUL.

endoderm (responsible for forming the lining of the gut and digestive glands). In amphibians, the three most common sites for malformations are the limbs, the jaws, and the eyes. In each of these regions, malformations are the result of mesodermal and/or ectodermal problems (i.e., legs, jaws, and eyes have no gut tissue and thus no endodermal contribution to their development). There can be no doubt that endodermal malformations arise, but because they are internal, these malformations are less likely to be observed. Additionally, because gut function is essential, animals with endodermal malformations may not survive their tadpole stage to reach metamorphosis, the age where malformations are detected, using current field collection protocols.

LIMB DEVELOPMENT

Perhaps no aspect of amphibian developmental biology is better known than the process of limb formation.[17] Not only does the study of amphibian limb development have a long history, but several high profile labs continue to study this important topic. To understand the malformed limbs of amphibians, it is useful to have in mind seven concepts underlying the development of normal limbs:

1. limbs develop from limb buds;

2. limbs have three axes: proximal–distal (from the body out to the fingers or toes); dorsal–ventral (back to belly); and anterior–posterior (head to tail);

3. limb buds give rise to limbs from proximal to distal, which means the farther a structure is from the body, the later it will develop;

4. limb components develop in the following order: cartilage, blood vessels, nerves, and muscles;

5. the shape of limbs varies by species (compare the long limbs of frogs to the short limbs of toads), and within an individual by position (compare forelimbs to hindlimbs) and side (limbs on opposite sides are mirror-image oriented);

6. limb growth after the shape pattern has been established is critical for proper limb function;

7. amphibian limbs, especially the developing limbs of anurans and larval and adult limbs of some salamanders, have impressive abilities to regenerate lost structures.

Some Details   As mentioned above, mesoderm and ectoderm combine to form limbs. To begin, aggregations of mesoderm from an area known as the lateral plate form beneath the surface ectoderm to create limb buds. At the distal end of each limb bud, the surface ectoderm thickens to form a discrete growth zone known as the apical ectodermal ridge (AER). Normal limb development is dependent on the interaction between limb bud mesoderm and the AER.

As the limb bud elongates, it contains an embryonic tissue termed mesenchyme. Mesenchyme consists of embryonic cells found as individuals or in groups, rather than in sheets. Mesenchymal cells can be derived from mesoderm or ectoderm (for example neural crest cells, which are mesenchymal and produce pigment, derive from ectoderm). Mesenchymal cells form precursors to cartilage (the formation of cartilage is called chondrogenesis) that will develop into the bones of the shoulder and pelvic girdles, and later into the bones of the limbs (the formation of bone is termed osteogenesis).

During early stages of outgrowth, the limb bud is invaded by structures called angiogenic cords (the primordia of blood vessels), which establish a blood supply (the formation of blood vessels is termed angiogenesis). Nerves follow and make motor and sensory connections (a process called neurogenesis). Finally, cells for the precursors to muscle (myogenic primordia) enter, condense, and form muscle masses, which subsequently segregate to form individual muscles (the process of developing muscle tissue is termed myogenesis).

Limbs develop not only by the assembly of structures but also by their disassembly. Areas of programmed cell death (necrotic zones) appear late in development, for example in the mesenchyme between the digits, wrists and ankles, elbows and knees, and armpits and groin. This process of cell death (called apoptosis) plays an important role in shaping limbs.

After the limb pattern is established, a period of rapid growth follows, during which muscles and skeletal tissues mature, tendons and ligaments are formed, and joints develop. The limb must be functionally innervated and moving for this to occur properly. As Lanyon and Rubin point out (p. 1):[16]

The development of an organism from an embryo into a normal adult is a complex process requiring both appropriate genetic instructions and adequate nutrition. However, this input alone is not sufficient to form a functional skeleton, since development occurring under these conditions (as in a paralyzed limb) produces skeletal structures that lack the detailed shape, mass, and arrangement of tissue necessary for load bearing. 'Normalcy' of architecture, and the structural competence that it reflects, is achieved and maintained only as a result of an adaptive response of the cells to load bearing. This response is functional adaptation.

During development, motor neurons supplying limb muscles are overproduced by anywhere from a factor of two to a factor of eight.[18] Motor neurons that "compete" successfully, live (i.e., make correct synaptic connections on myotubes during the development of limb reflex movements); those that do not "compete" successfully, die. While the absence of a limb does not affect motor neuron proliferation, migration, and initial differentiation, limb absence results in the death of nearly all motor neurons responsible for its innervation. Somatosensory neurons that would normally provide touch, pressure, pain, and temperature information from the limbs are similarly affected. It has long been known that the effects of developing limb amputation depend on developmental stage. In bullfrogs (*Rana catesbeiana*), amputation of a late-stage tadpole limb (when reflexes have become established) results in a degeneration of motor neurons at metamorphosis.[19] After metamorphosis, however, motor neurons do not degenerate following amputation.[18] This developmental process can assist in interpreting causes of missing limbs; under certain circumstances the absence of motor neurons distinguishes limbs never present from limbs once present and now absent.

In vertebrates, limb abnormalities and dysfunctions are among the most commonly encountered developmental problems. In part this is because developing limbs are more or less self-contained: limb buds either have what it takes to make an intact functioning limb or they do not, and when they do not, adjacent or interacting tissues or organs cannot assist by taking corrective action. Further, while limbs are critical for proper behavior and for ecological viability, they are not necessary for basic body functions. Therefore, unlike developmental problems in vital organs or organ systems, glitches in limb development tend not to be fatal. For frogs, this point is particularly important, because aquatic, swimming tadpoles do not need functioning limbs to survive, or even to thrive. Forelimbs are so unimportant to tadpoles that they are tucked under gill covers. Therefore, as tadpoles, frogs with malformed limbs may be behaviorally and ecologically unaffected. At metamorphosis, of course, this situation changes abruptly from being simply curious to us to being vitally important to the frog.

During metamorphosis, limb assembly and coordination occur quickly, because the requirement to have four functioning limbs wired correctly and working together is behaviorally and ecologically essential. Metamorphosis is the most vulnerable time in a pond-breeding amphibian's life history.[20] Metamorphosing amphibians are concentrated along wetland edges and relatively immobile; they are easy prey for garter snakes, raccoons, opossums, and various bird species. And whereas having limbs that work is not a requirement for amphibian life, having working limbs at metamorphosis enables amphibians to escape from predators, and is therefore crucial for amphibian survival.

JAW DEVELOPMENT

Vertebrate jaws develop from an embryonic structure called the first visceral arch.[21] Visceral arches (there are six total) form in the region of the pharynx (neck) and are produced from all three germ layers (ectoderm, mesoderm, and endoderm). Visceral arches form the structures responsible for swallowing and breathing (including the formation of gills in fishes and aquatic amphibians). The major vessels of the circulatory system, including the heart, aorta, and big vessels to and from the arms and brains, develop in visceral arches and later migrate into the thorax.

To form jaws, the neural crest-derived mesenchyme of each side of the first visceral arch splits in two to form the upper jaw (mandibular process) and the lower jaw (maxillary process). These processes grow towards the front of the face, meet, and fuse. At metamorphosis, frog jaws undergo further change as they reconfigure from the small scraping, sucking cartilaginous mouthparts of tadpoles, to the large grabbing, biting, bony mouthparts of adults.

EYE DEVELOPMENT

In vertebrate animals, proper eye development is the result of a series of complicated ectodermal and mesodermal tissue interactions.[21] In amphibians, these interactions also include gut (endodermal) tissue—because during swallowing, amphibian eyes close, retract, and push food down their throats—making these processes even more complex.

Eye development in vertebrates begins with the brain, specifically the forebrain (which also produces the cerebral cortex). Optic vesicles form on either side of the forebrain and grow out to where the eyes will eventually be positioned. When an optic vesicle meets the surface ectoderm, the ectoderm becomes transformed to create a lens placode, which invaginates to form a lens vesicle. This lens vesicle breaks away from the surface ectoderm, and this ectoderm becomes a portion of the cornea. At the same time, the lateral surface of the optic vesicle (remember, this is the extension of the forebrain) invaginates to form a nearly complete, bilayered optic cup. The inner surface of the optic cup gives rise to the neural portion of the retina; the outer layer of the optic cup gives rise to the pigmented epithelium of the retina. The optic stalk (now connecting the optic cup to the forebrain) persists, and guides axons from retinal ganglion cells back to the developing brain. When your family physician or ophthalmologist examines your retina, he or she is peeking at a derivative of the same brain structure that created your mind.

In the eye, the iris develops from cells that originated both from the pigmented epithelium and neural crest (ectoderm). Connective tissues in the eye and outside of the eyeball derive from neural crest and mesoderm. Blood supply comes from vessels associated with the first aortic arch (the same source as the blood supply to the jaws). Muscles that move the eyes, including in amphibians the levator and retractor bulbi muscles that assist food manipulation and swallowing, develop from head mesodermal tissue.

TIMEFRAMES AND THE APPEARANCE OF MALFORMATIONS DURING DEVELOPMENT

Tadpoles typically hatch from eggs and metamorphose into terrestrial juveniles within a single season, although the tadpoles of larger aquatic frogs such as bullfrogs, green frogs (*Rana clamitans*), and mink frogs may overwinter at least once.[22] In amphibians, rates of development are temperature dependent, so within a species, animals in northern populations take longer to complete life cycles than animals in southern populations. Similarly, animals in spring-fed wetlands grow more slowly than animals in pothole wetlands.

Rates of development vary across species, from as little as four weeks among treefrogs in the family Hylidae, (such as cricket frogs [*Acris crepitans*], eastern gray treefrogs [*Hyla versicolor*], and chorus frogs [*Pseudacris triseri-ata*]) and true toads in the family Bufonidae (such as American toads [*Bufo americanus*]), to over seven weeks (true frogs in the family Ranidae, such as northern leopard frogs). Some tadpole stages last much longer, including those of the bullfrogs, green frogs, and mink frogs mentioned above.

Malformed structures can only be observed when, or after, structures have (or would have) formed, so different types of malformations arise at different times during development. Eyes form embryonically, so gross eye malformations appear quickly and will be visible in both tadpoles and in adults. Jaws also form early in development, but unlike eyes, get completely reworked at metamorphosis. Gross jaw malformations can be seen in tadpoles or in adults, although they are often most conspicuous in adults. Limbs form in late-stage tadpoles, and limb malformations are often not apparent until tadpoles reach pre-metamorphic developmental stages. In species with a short tadpole stage, such as American toads, limb formation may precede metamorphosis by as little as one or two weeks.

TWO

MALFORMED FROG TYPES

Die Anatomie ist das Schicksal.
(Anatomy is destiny.)

SIGMUND FREUD[1]

In an attempt to organize, and to correlate effect with cause, malformations, whether human, frog, or otherwise, have tended to be divided into types. Malformation types are almost always based on: (1) structures absent or reduced, (2) structures duplicated (or multiplied), and (3) structures present but otherwise abnormal (e.g., eye position, jaw shape, limb shape, skin color, pigment pattern). Definitions of malformed frogs have been provided by Carol Meteyer and her colleagues.[2] The team we assembled in 2001 to re-examine the hottest of the Minnesota malformed frog hotspots slightly modified this terminology,[3] which is in turn slightly modified and presented in Table 2.1. Note that internal soft tissue features, such as malformed gonads and larynxes (voice box), which are known to occur following exposure to some environmental contaminants (see Chapter 4), and oral deformities in marginal papillae, tooth rows, and jaw sheaths caused by the *Batrachochytrium dendrobatidis* (chytrid) infections[4] are not included here.

WHICH ANIMALS TO REPRESENT

Collections of malformed frogs have never been done in a sufficiently systematic way to permit meaningful statistical analyses (see discussion in Chapter 5). About all you can really say is that animals of species X with particular malformation Y were collected at wetland Z on day A. Absence of malformation types in collections may or may not reflect

TABLE 2.1.   *Classification and definitions of frog abnormalities*

Type and Location of Abnormality	Description
Craniofacial	
Anophthalmia	small eye
Brachygnathia	abnormal shortness of lower jaw
Eye discoloration	iris pigment discolored or missing
Eye displacement	eye displaced laterally, medially, cranially, or caudally
Microcephaly	blunt nose; shortened upper jaw
Microphthalmia	missing eye
Forelimb and Hindlimb	
Abnormal pigment	pigment pattern missing or abnormal
Amelia	missing limb
Taumelia	long bone bent back on itself forming > 90° angle
Bony expansions	distal end of a bone expands into spongy balloon
Brachydactyly	normal number of metatarsals but abnormal number of phalanges
Curved long bone	all long bones bend at the site of artery penetration
Ectrodactyly	complete absence of digit including metatarsal bone
Ectromelia	missing limb segments (e.g., femur present but rest of limb missing)
Hemimelia	shortened bone
Micromelia	entire limb present but all limb elements shortened
Polydactyly	complete extra digit including metatarsal bone
Polymelia	complete extra limb
Skin webbing	band of skin crossing a joint
Syndactyly	fusion of digits
Whole Body	
Bloated body	swelling in the torso and limbs of the animal

Note: Data from Meteyer and colleagues.[2]

their absence in the sampled wetland. Percentages of malformed frogs collected as newly metamorphosed animals (which can and do vary from day to day) may or may not reflect true percentages. Further, percentages of malformed metamorphosed frogs may or may not represent percentages of malformed tadpoles or malformed embryos in that population.

What all this really means is that either way radiographs are represented—as a full collection (prohibitively costly, not to mention redundant) or as a highly selected subsample—there is bias. And bias is bias—no matter whether there is a little of it or a lot of it—and to be honest we have to acknowledge it and deal with it.

As mentioned above, frog malformations can be categorized simply as parts missing, parts present but abnormal, and parts extra, and that is how I organized this chapter. For each category, the best representatives were chosen. Radiographs were selected that would allow comparisons[5] of species, sites, and malformation types. In all, 63 radiographs representing eight species are presented. Many specimens used here were collected from Minnesota by Minnesota Pollution Control Agency field biologists and by Dave Hoppe. Several specimens were collected at one Wisconsin site by Dan Sutherland. Specimens were collected at two California sites by Pieter Johnson, at one southeastern Indiana site by Charles Facemire, at two east-central Indiana sites by Laura Blackburn, and at a National Wildlife Refuge in Alaska by Mari Reeves.[6] An historic specimen, collected in Ohio and housed in the Hefner Zoology Museum at Miami University in Oxford, Ohio, was sent to me by Curator Mike Wright. A specimen in the collection at the Field Museum in Chicago was sent to me by Alan Resetar.

UNILATERAL AMELIA

TOO FEW HINDLIMBS

Example 1 Each malformed frog has its own story to tell. The details of the morphologies we see reflect not only the malformation, and perhaps the cause of the malformation, but also how the animal responded to the malformation. In Figure 2.1 the malformation is a missing right hindlimb.[7] Notice that not only is the hindlimb absent, but the ilium—the lateral pelvic element—on the same side is also missing (agenesis in the terminology of Meteyer and her colleagues[2]). This animal shows no signs of scarring, which is not surprising—imagine the violence of a failed predation attempt that not only removed a limb, but also removed a portion of the pelvis. Such trauma is inconsistent with life. A bite that removes a leg would also tear the large-bore femoral artery supplying blood to the leg, and likely result in the animal bleeding to death (the human circulatory system is organized in the same way and ER docs and battlefield medics know this injury and fear for the patient). The malformation illustrated here is not due to failed predation, but instead must be due to developmental error; an error that not only resulted in the faulty initiation of limb formation, but also resulted in the failure of a portion of the associated pelvis to develop.

The drive to survive is strong. Note that on the opposite side of the malformation, the bony complex of the ilium, ischium, and femur are

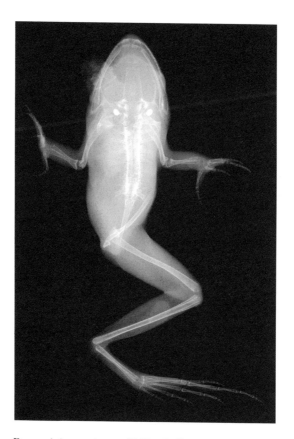

FIGURE 2.1 *Rana pipiens.* 36 mm SUL. Collected on 11 September 1998 at the ROI site in Meeker County by Minnesota Pollution Control Agency field biologists. The animal shown in Figure 2.22D is also from ROI. See Chapter 3 for a description and Figure 3.3 for a photograph of this site.

displaced laterally, towards the side of the missing limb, making the articulation with the intact leg positioned on the opposite side of the midline coccyx. The articulation between the ilium and the transverse processes of the sacrum (the sacroiliac joint) appears to be intact, and so skeletal disarticulation, in the formal sense, did not occur. Note also the curvature (scoliosis) of the vertebral column, with the convexity towards the unaffected side. This combination of displaced pelvis and spinal curvature serves to place the intact hindlimb in a position that is more (about 28°) behind than beside the animal, where single-limbed propulsion can be more effective. Unlike the missing limb and ilium on the opposite side, these abnormalities of alignment are not the result of development gone wrong, but instead are secondary effects, reflecting a remodeling of the position of the skeletal elements following use.

Example 2   Similar malformations are seen in the same species from different sites. This northern leopard frog (Fig. 2.2) collected during the summer of 1997 from a lesser-known Minnesota hotspot exhibits a hindlimb malformation nearly identical to the northern leopard frog shown in Figure 2.1.

According to Meteyer et al. (p. 155),[2] amelic frogs comprised 3.9% (22 of 570) of all malformations documented during 1997 and 1998 field surveys at Minnesota sites. Frogs from five of eight Minnesota sites exhibited amelia; at two of these sites, amelia was seen in both years.

Meteyer and her colleagues also point out that 72% of amelic frogs in their Minnesota study also presented with malformations of at least one ipsilateral pelvic element, which included absence of the ilium, ischium, and pubis on the side with the missing limb.

Example 3   The same malformation types can be seen in different species. As with the northern leopard frogs illustrated in Figures 2.1 and 2.2, the bullfrog illustrated in Figure 2.3 is also completely missing a hindlimb as well as an ilium (the radio-dense blobs where you would expect to find the ilium are likely ingested soil and sand particles in the intestine). Unlike the northern leopard frog examples, however, the spine and pelvis of this bullfrog are properly aligned. One explanation (although such posthoc analyses are often, and perhaps usually, incorrect) is that bullfrog hindlimbs do not exhibit the same degree of remodeling because unlike newly metamorphosed northern leopard frogs, (which are terrestrial), newly metamorphosed

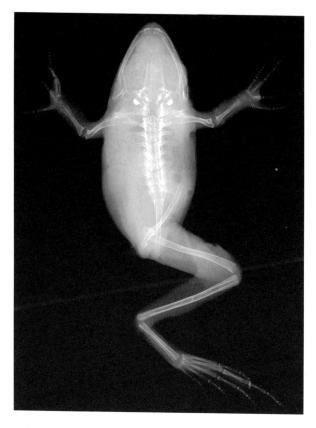

FIGURE 2.2 *Rana pipiens.* 36 mm SUL. Collected on 8 August 1997 from the WIN site by Minnesota Pollution Control Agency field biologists. The animals shown in Figures 2.12, 2.23, and 2.53 are also from this site.

bullfrogs tend to be aquatic and face less resistance during limb propulsion in water. Note also that the intact ilium is malformed; rather than being curvilinear, this bone consists of straight segments with sharp bends.

Example 4   A northern leopard frog tadpole at about Gosner stage 41 (Fig. 2.4) demonstrates developing amelia and an associated missing ilium. Comparing this radiograph to those in the developmental sequence illustrated in Figure 1.3 suggests that even at this early stage, the intact ilium

FIGURE 2.3   *Rana catesbeiana*. 42 mm SUL. Collected in California during the 1997–1998 field season by Pieter Johnson.[3] See Figures 2.29B, 2.32, 2.35, 2.51 and 3.15 for images of other animals from this site.

is oriented more mediolaterally (side to side) than rostrocaudally (head to pelvis), an orientation that would favor propulsion in the remaining limb. It would be interesting to know whether this morphology, in such a young animal, is due to remodeling following use (again, hindlimbs are not functionally as important in tadpoles as they will become in adults) or whether during the process of development this "normal" limb formed in an abnormal orientation. One only has to consider nipples in human males to know that not all structures form for purely functional reasons.

   This animal was collected by the author at Welch Lake, a natural wetland being used for aquaculture—to raise walleye (*Stizostedion vitreum*) fry to fingerling size before stocking them in the large recreation lakes of the Okoboji region in northwestern Iowa. This form of husbandry involves the applications of chemical piscicides and herbicides. For an account of the impact of such activities on the amphibians of the region see Lannoo.[8]

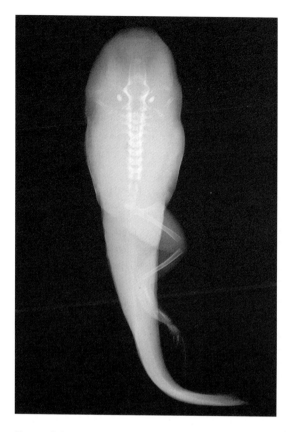

FIGURE 2.4 *Rana pipiens.* 32 mm SUL. Ectromelic tadpole collected by the author at the outlet of Welch Lake, located in Dickinson County, in northwestern Iowa. The animal shown in Figure 2.8 is also from Welch Lake. See Chapter 3 for a description and Figure 3.12 for a photograph of this site.

Example 5   This unusual northern leopard frog (Fig. 2.5) is postmetamorphic (note rounded snout, high level of ossification, and well-developed forelimbs) but retains a full tail. It is amelic, with a missing ilium on the affected side. It is also ectromelic, with missing limb segments below the knee joint on the limb that is present. This femur exhibits an "abrupt termination" in the terminology of Meteyer and her colleagues;[2] it may also be hemimelic (shortened), although without a normal limb for comparison this is difficult to determine.

This frog was collected at an odd Wisconsin site. According to a local fisherman, this pond was normal until it and the adjacent farmland were bought by a poultry processing business. They began using the fields to spread chicken manure and that is when he began noticing malformations. Such anecdotal evidence is characteristic of malformed frog sites, because whereas few people pay attention to normal, day-to-day occurrences, novelty attracts. It is also true that few researchers have the

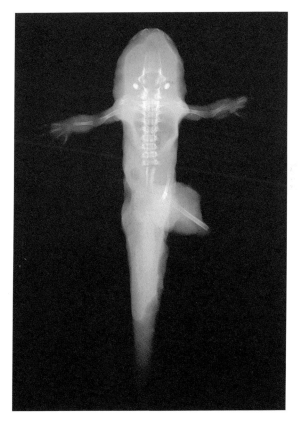

FIGURE 2.5 *Rana pipiens.* ~32 mm SUL. Ectromelic animal collected by Dan Sutherland from a farm pond near Arcadia, in Trempealeau County, Wisconsin, in 1998. Sutherland did not find metacercariae of the trematode *Ribeiroia ondatrae* in this frog. Frogs shown in Figures 2.7, 2.11, 2.13, 2.19, 2.20, 2.22A, B, C, 2.25, and 3.13 represent additional animals collected at this site.

resources to do exhaustive (and expensive) water chemistry analyses to search for potential chemical causes.

Example 6   This green frog (Fig. 2.6) is not only missing a hindlimb, but is also missing a portion of its ilium associated with the missing limb. The rostral (headward) portion of the ilium is ossified, but you can see that it is thin, short and exhibits a concave lateral surface—the opposite of what we see in normal ilia. The tissue and displaced arm on the left side of the radiograph are dissection artifacts.

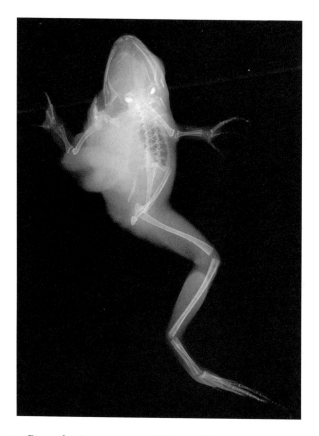

FIGURE 2.6   *Rana clamitans.* 35 mm SUL. Collected by the author on 22 September 1998 at a site south of Greenfield, in Hancock County, Indiana.

The wetland where this frog was collected had been transformed into a lake, around which a subdivision was being built. Previously it had been a wet woodlot, untillable, where farm equipment and supplies, including agricultural pesticide containers, had been discarded (again, note the anecdotal nature of this information).

Example 7   This northern leopard frog (Fig. 2.7) is missing a leg but its pelvis is intact. Note the kink in the contralateral femur at about one

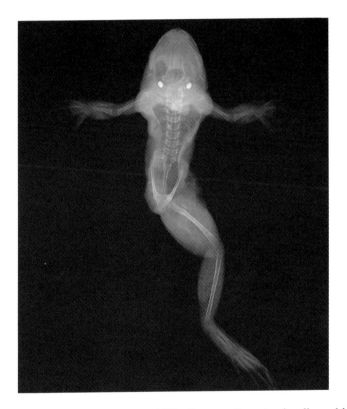

FIGURE 2.7   *Rana pipiens.* 31 mm SUL. Ectromelic animal collected by Dan Sutherland in 1998 from a farm pond near Arcadia, in Trempealeau County, Wisconsin. Sutherland did not find metacercariae of the trematode parasite *Ribeiroia ondatrae* in this frog. Frogs shown in Figures 2.5, 2.11, 2.13, 2.19, 2.20, 2.22A, B, C, 2.25 and 3.13 represent additional animals collected at this site.

FIGURE 2.8   *Rana pipiens.* 37 mm SUL. Amelic juvenile collected by the author at the outlet of Welch Lake, located in Dickinson County, in northwestern Iowa. The animal shown in Figure 2.4 is also from Welch Lake. See Chapter 3 for a description and Figure 3.12 for a photograph of this site.

quarter of the length down the shaft. The soft tissue exhibits a varying opacity because this animal was dissected to determine the presence of parasites.

Example 8   This northern leopard (Fig. 2.8) frog is amelic and exhibits a complete but displaced pelvis. Not only is the position of the pelvis asymmetrical, but also bone development is asymmetrical. The ilium on the affected side is thicker and perhaps bifurcated towards the sacrum.

This animal was collected at the Welch Lake site in Iowa, which is heavily impacted by aquaculture practices.[8]

TOO FEW FORELIMBS

Example 9   Rarely, animals with affected forelimbs have been found.[9] For example, a population of *Rana ornativentralis* from Japan was discovered that included individuals with supernumerary forelimbs.[10]

The northern leopard frog illustrated in Figure 2.9 is not only missing a forelimb, but is also missing its entire shoulder girdle, strongly suggesting that this malformation represents a developmental problem. It is unknown why either hindlimbs or forelimbs are selectively affected, but it is known that genes such as Hoxc-6 and Tbx-5 are differentially expressed in forelimbs, while Hoxc-10 and Tbx-4 are differentially expressed in hindlimbs.[11] Therefore, alterations to particular genes or gene products could favor either forelimb or hindlimb effects.

A handful of authors have assumed that because tadpole forelimbs develop in the branchial cavity, they are protected from chemical insults. In part this may be due to the mistaken impression that forelimbs develop

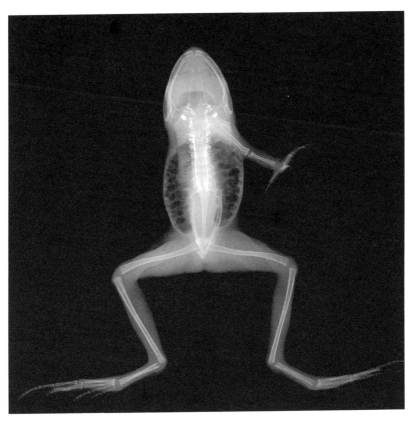

FIGURE 2.9 *Rana pipiens.* 38 mm SUL. Collected on 11 August 1997 from the OSP site by Minnesota Pollution Control Agency field biologists.

"in a fluid-filled sac."[12] In fact, forelimbs develop in the gill (branchial) chambers, which are open to the environment. Gills and, by virtue of their location forelimbs, are irrigated with pond water as tadpoles buccal pump (i.e., move their mouths to bring in water, which passes over the gills to oxygenate blood). In fish, water brought in during buccal pumping is passed out the gill (opercular) flaps on both sides. In tadpoles, the opercular flaps are closed, and water passes out a tube called the spiracle, which in most North American frogs is located on the left side of the body. It is possible that in a buccal pumping tadpole at rest, the forelimbs, and especially the left forelimb, are more exposed to chemical insults than are the passively dragging hindlimbs. In truth, tadpoles are so permeable that all structures, no matter whether they are internal or external, are likely to be about equally exposed.

FIGURE 2.10   *Rana septentrionalis.* 38 mm SUL. Collected by Dave Hoppe in 1999 at the CWB site in Crow Wing County, Minnesota (specimen # CWB7168018M). Frogs shown in Figures 2.26, 2.30, 2.33, 2.36D, 2.38, 2.39, 2.41, 2.42, 2.43, 2.44 and 2.46 are also from CWB. See Chapter 3 for a description and Figure 3.5 for a photograph of this site.

Example 10 As with hindlimb malformations, we observe the same fore-limb malformations across species. Similar to the northern leopard frog shown in Figure 2.9, this mink frog (Fig. 2.10) is not only missing a fore-limb, but most of the associated shoulder girdle is either absent or incom-pletely ossified. This animal was not scarred. In combination with the poorly developed ossification (indicating that this is a newly metamorphosed ani-mal and therefore that the forelimb had been recently exposed), the absence of scarring strongly suggests that this malformation represents a develop-mental problem.

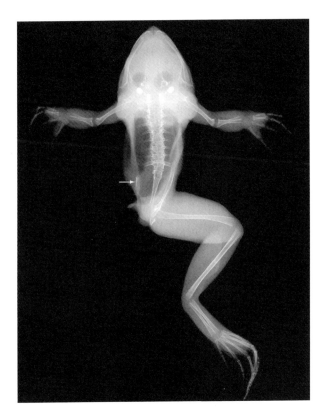

FIGURE 2.11 *Rana pipiens.* 33 mm SUL. Collected by Dan Sutherland in 1998 from his Trempealeau County, Wisconsin, site. Arrow indicates slight ossification of the affected ilium. Frogs shown in Figures 2.5, 2.7, 2.13, 2.19, 2.20, 2.22A, B, C, 2.25 and 3.13 represent additional animals collected at this site.

Example 11   Without question the most common malformation type we observe is missing hindlimb segments (ectromelia). This northern leopard frog (Fig. 2.11) exhibits a small, soft tissue growth in the region of the missing limb, and a small degree of ossification in the rostral (headward) portion of the associated ilium.

Example 12   Unlike most of the missing limbed (amelic) animals previously shown, the animal shown in Figure 2.12 with missing limb segments (ectromelia) has an intact and properly aligned pelvis and a developed portion of its hindlimb. This morphology indicates limb initiation had begun, but subsequent growth and pattern formation were affected. There appears to be a slight scoliosis (about 9°) favoring limb thrust from behind the frog.

Note the bone in the abnormal limb. This affected femur is twice as thick as the normally sized femur on the contralateral side, and the bony tissue appears less dense. Rather than having the usual organization of compact bone surrounding a cancellous center, the affected femur presents as though its compact and cancellous bone are mixed (or perhaps there are different densities of cancellous bone). Meteyer and her colleagues[2] term this a "bulbous expansion" though none of their examples exhibit this level of involvement. Look closely at the region near the pelvis and note that this femur has a normal-appearing proximal portion, although it does not contact the pelvis. As with the amelic animal described previously, there is no evidence of scarring. This is a developmental malformation and whatever went wrong with this limb likely occurred immediately following limb bud outgrowth.

Meteyer and her colleagues[2] also point out that animals with missing limb segments were the most common malformation type (200 of 570; 35%) found in their study, and that these animals were found at all of the Minnesota study sites in each of the two years of the study (p. 163). Further, they found that limb truncations at various levels comprised nearly one-half of their observed malformations (p. 169).

Example 13   Whereas missing limb segments comprise a universally recognized limb malformation type in humans as well as amphibians, radiographic profiles of frogs with missing limb segments vary. When examined

FIGURE 2.12   *Rana pipiens.* 50 mm SUL. Collected 15 August 1997 from the WIN site by Minnesota Pollution Control Agency field biologists. According to Sutherland, these frogs did not have metacercarial cysts from the trematode *Ribeiroia ondatrae.* The animals shown in Figures 2.2, 2.23, and 2.53 are also from this site.

grossly, this northern leopard frog (Fig. 2.13) from Trempealeau County, Wisconsin, resembles the frog shown in Figure 2.12: the axial skeleton is intact (here with few signs of scoliosis) and one leg is missing at the level of the mid-femur. But note how different the femurs of these two animals appear when viewed radiographically. Unlike the femur of the animal shown in Figure 2.12, this femur does not expand into a well-defined region of ill-defined bone; instead the compact bone thickens and expands away from the shaft to form half of a sphere—a bony "bubble"—and the cancellous bone within this region is less dense than bone closer to the

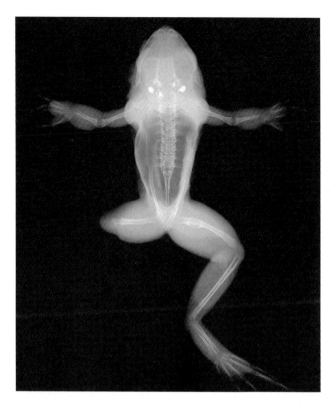

FIGURE 2.13   *Rana pipiens.* 35 mm SUL. Collected by Dan Sutherland in 1998 from his Trempealeau County, Wisconsin, site. According to Sutherland, these frogs did not have metacercarial cysts from the trematode *Ribeiroia ondatrae.* Frogs shown in Figures 2.5, 2.7, 2.11, 2.19, 2.20, 2.22A, B, C, 2.25 and 3.13 represent additional animals collected at this site.

body. Where the leg is absent, both the femoral shaft and the spherical expansion end; the bony sphere never fully forms. At its termination, the shaft shows minor, irregular bone growth, and perhaps an additional distal bony element. Meteyer and her colleagues[2] define this condition (p. 161) as "termination with cortical separation" and further describe this morphology as "a delicate radiodense ring formed around the distal cortex of the affected femur."

Just as interesting is an examination of the "good" femur: the bone on the opposite side. Although the limb appears normal in life, radiographs show that the femur of this limb is also affected. Note the presence of a bony sphere similar to the one at the site of the malformation on the opposite femur. This sphere, closer to the body than its counterpart on the malformed limb, produces a slight kink in the bone, but otherwise the limb develops just about normally. Meteyer and her colleagues[2] never observed animals with this condition in legs that otherwise developed normally (a condition that could be called "cortical separation without termination").

Most of the 28 animals collected from this site showed some form of compact bone expansion forming a "bubble" (see Trempealeau County, Wisconsin, site description in Chapter 3). This malformation type has also been observed in Blanchard's cricket frogs (*Acris crepitans blanchardi*) collected from east-central Indiana (Fig. 2.14).

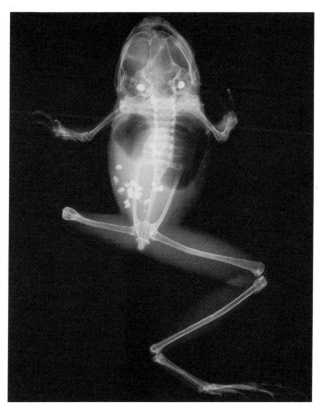

FIGURE 2.14   *Acris crepitans.* 19 mm SUL. Newly metamorphosed animal collected in 1999 from Mary Gray Bird Sanctuary near Connersville, in Fayette County, Indiana, by Laura Blackburn.

Example 14   An ectromelic Blanchard's cricket frog (*Acris crepitans blanchardi*; Fig. 2.14), collected from east-central Indiana, exhibiting a pattern of bone expansion at the point of limb loss similar to the northern leopard frog shown in Figure 2.13. This animal was collected at a nature preserve where few malformed frogs were found. Other malformed frogs collected at this site did not exhibit this pattern of bone expansion.

When considered together with the northern leopard frog from Wisconsin illustrated in Figure 2.13, this cricket frog from Indiana reinforces the notions that the same malformation type can be found in different species (from different taxonomic families) and from geographically distant sites.

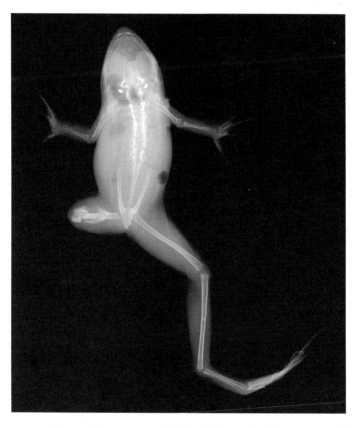

FIGURE 2.15   *Rana pipiens.* 46 mm SUL. Collected 8 August 1997 from the FGC site by Minnesota Pollution Control Agency field biologists.

Example 15   This northern leopard frog (Fig. 2.15) exhibits a pattern of bone expansion intermediate between the leopard frogs illustrated in Figures 2.12 and 2.13. The expansion is broad, encompasses most of the femur, and is as cancellous (spongiform) as the animal shown in Figure 2.12. But also note the bubble-like appearance characteristic of the bone of the animal shown in Figure 2.13.

Example 16   At first glance the femur of this northern leopard frog (Fig. 2.16) just ends—an abrupt termination in the morphology of Meteyer and her colleagues,[2] and a morphology reminiscent of amputations. However,

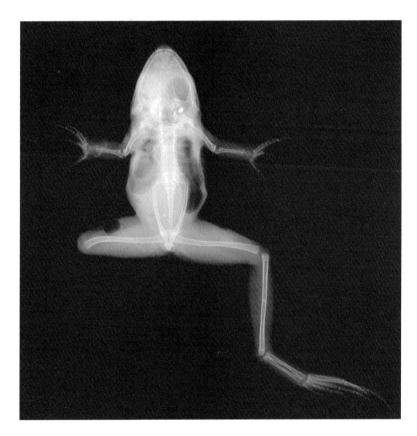

FIGURE 2.16   *Rana pipiens.* 40 mm SUL. Collected 24 September 1997 at the SUN site, in Ottertail County, Minnesota, by Minnesota Pollution Control Agency field biologists. Animals from this site are also shown in Figures 2.54 and Color Plate 4.

a close examination of the distal end of this bone reveals a curvature not present in the femur of the intact contralateral side, which suggests a developmental origin. Why would an amputation at the knee produce curvature in the femur?

Example 17 This northern leopard frog (Fig. 2.17) exhibits hemimelia (shortening) of the tibiofibula with a reduction in numbers and size of foot elements distally. Note the cancellous nature of the abnormal tibiofibula, and how the ends of both bones are distinct from the central

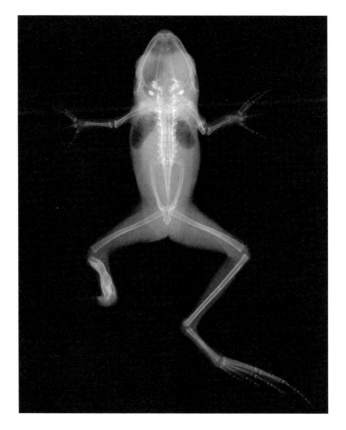

FIGURE 2.17 *Rana pipiens.* Collected 12 August 1997 from the HYD site, in Polk County, Minnesota, by Minnesota Pollution Control Agency field biologists. See Chapter 3 for a description and Figure 3.6 for a photograph of this site.

mass, which appears divided along a plane perpendicular to the long axis of this complex. In the frogs observed by Meteyer and her colleagues,[2] bulbous terminations were only associated with the femur. But as they point out (p. 170), despite examining a large number of animals (570 frogs) Meteyer's group likely did not observe the full range of realized malformation types (and neither have I).

Example 18   A recently metamorphosed American toad from the Chicago area (Fig. 2.18) has a missing foot and a foreshortened and thickened tibiofibula. This is one of the least photogenic animals in my assemblage, but it is included here because although foreshortened femurs are a common malformation type, foreshortened tibiofibulas are not.

Example 19   This northern leopard frog (Fig. 2.19) exhibits a combination of malformations associated with hindlimb formation and tail resorption.

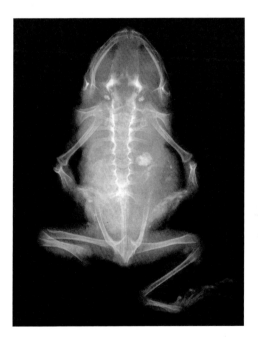

FIGURE 2.18   *Bufo americanus.* 22 mm SUL. Collected in Cook County, Illinois, and sent to the author by Alan Resetar (specimen # FMNH 256174).

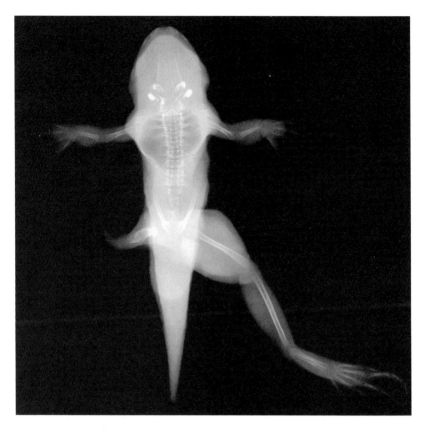

FIGURE 2.19   *Rana pipiens.* ~29 mm SUL. Collected in 1998 by Dan Sutherland from a farm pond in Trempealeau County, Wisconsin. Sutherland reported these frogs did not have *Ribeiroia ondatrae* metacercariae. Frogs shown in Figures 2.5, 2.7, 2.11, 2.13, 2.20, 2.22A, B, C, 2.25 and 3.13 represent additional animals collected at this site.

Note the shadow of the unresorbed tail, and a mass associated with the base of the tail. The ectromelic limb has a foreshortened femur with an expanded, cancellous, distal end. Beyond this are cartilaginous and soft tissue elements.

Example 20   Field reports are usually accurate but can be misleading. This northern leopard frog (Fig. 2.20) was initially reported as exhibiting "ectromelia with hemimely of the tibiofibula." A close look at the "normal" hindlimb, however, reveals a bone expansion above the knee associated with an early termination of the femur and an apparent proximal expansion of the tibiofibula that curves toward the femur. Anatomically, it is difficult to determine where the knee is located in this limb. Although it is technically correct to describe this malformation as "cortical separation without termination" (see Meteyer et al.[2] and Fig. 2.13), there is clearly something beyond normal cortical separation occurring here.

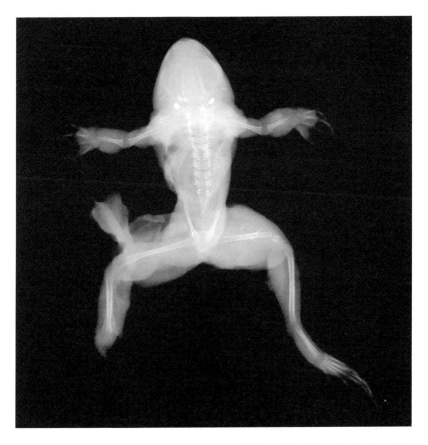

FIGURE 2.20 *Rana pipiens.* 32 mm SUL. Collected by Dan Sutherland in 1998 from a farm pond in Trempealeau County, Wisconsin. Sutherland reported these frogs did not harbor metacercarial cysts of *Ribeiroia ondatrae.* Frogs shown in Figures 2.5, 2.7, 2.11, 2.13, 2.19, 2.22A, B, C, 2.25 and 3.13 represent additional animals collected at this site.

The ectromelic limb is also noteworthy. Meteyer and her colleagues[2] noted termination with cortical separation associated with the femur only; here, this condition affects the tibiofibula. This is nothing more than a numbers issue, as Meteyer and her colleagues fully recognize. I wonder how many malformation types we have yet to discover.

Example 21   In the field, the loss of limb segments (ectromelia) at the same place, or roughly the same place, on both hindlimbs is a rare finding (see p. 167 in Meteyer et al.[2]). The animal shown in Figure 2.21 has a normal axial skeleton (the conversion from cartilage to bone is incomplete and so the contrast between bone and non-bony tissue is light). However, each hindlimb terminates at about the mid-femoral level and the ends of both femurs show club-like expansions consisting of light-density bone. Unlike several of the animals shown previously, these expansions do not en-

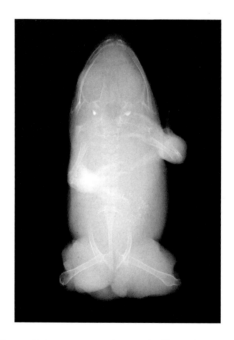

FIGURE 2.21   *Rana pipiens.* 46 mm SUL. Collected in 1999 by Dave Hoppe from Stevens County, Minnesota (specimen # STA7176078L). The bones of this animal are lightly ossified.

compass the normal shaft of the femur; they constitute the normal shaft of the femur. The tip of the femur on the left side of the radiograph has a compact bony cap; the tip of the femur on the right side has an irregular, small expansion. These terminal bony expansions contrast deeply with the radiographic appearance of bilaterally ectromelic limbs produced by exposure to UV-B,[13] which consistently have no bony expansions.

Meteyer and her colleagues[2] point out that in their 1997 and 1998 samples from Minnesota, lesions were symmetrical in only 8 of the 22 frogs with bilateral malformations.

Example 22   Other examples of bilateral ectromelia in northern leopard frogs are illustrated here (Fig. 2.22). In A, C, and D the ectromelia is not as symmetrical as the example shown in Figure 2.21. In A the femur and opposite tibiofibula are involved. In B the tibiofibulas are involved symmetrically. In C the tibiofibulas are involved asymmetrically. Note that in A, B, and C the affected bones have cancellous (spongy) expansions and terminal or near-terminal bony expansions characteristic of animals from this Wisconsin site. In D, the hemimelic (shortened) femur ends without expansion; the opposite limb exhibits a combination of ectromelic and misshapen elements, including cancellous bone expansions and a taumelia (bony triangle).

OTHER HINDLIMB PROBLEMS

UNILATERAL MICROMELY

Example 23   A field examination of this northern leopard frog (Fig. 2.23) from Minnesota characterized this animal as having hindlimb micromely (a small limb). Radiography reveals that this characterization is certainly correct although incomplete: there are several additional problems in this interesting animal. Primary defects include a kinked femur in the micromelic limb and an extra bony element near the junction of the affected limb with the pelvis. This extra element is either a femur or an ilium, and is also kinked. Pelvic elements are disarticulated at the level of the hindlimb junction; in particular the ilium on the affected side does not contact ischial and pelvic elements and may be foreshortened. Secondary effects include scoliosis and a shift in the position of the normal ilium toward the affected side; a morphology that may favor forward thrust by the normal limb.

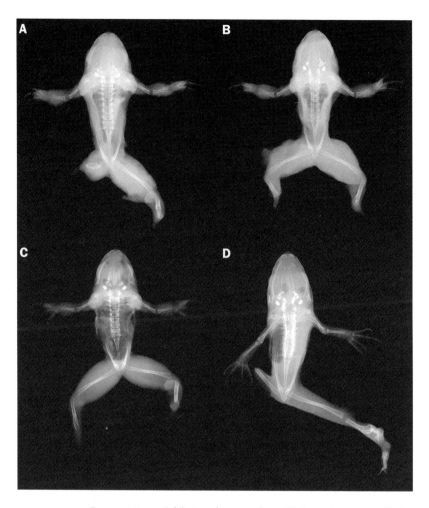

FIGURE 2.22   *Rana pipiens.* Additional examples of bilateral ectromelia in
northern leopard frogs. A, B and C were found in Dan Sutherland's
Trempealeau County, Wisconsin, site in 1998; D was collected from
ROI on 3 August 1998 by Minnesota Pollution Control Agency field
biologists. (A) 32 mm SUL; (B) 33 mm SUL; (C) 33 mm SUL; (D) 33
mm SUL. Frogs shown in Figures 2.5, 2.7, 2.11, 2.13, 2.19, 2.20, 2.25
and 3.13 represent additional animals collected at Sutherland's Wisconsin site. The animal shown in Figure 2.1 is also from ROI. See Chapter
3 for a description and Figure 3.3 for a photograph of the ROI site.

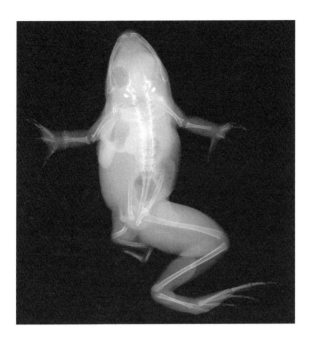

FIGURE 2.23 *Rana pipiens.* 40 mm SUL. Collected on 5 August 1997 from the WIN site by Minnesota Pollution Control Agency field biologists. The animals shown in Figures 2.2, 2.12, and 2.53 are also from this site.

Meteyer and her colleagues[2] observed that complete but malformed limbs comprised 18.4% (105 of 570) of all malformations documented during their 1997 and 1998 field surveys of Minnesota sites.

Example 24 These northern leopard frogs (Fig. 2.24), from separate sites in different ecoregions (NEY, in LeSueur County, is in the eastern deciduous forest of the Minnesota River Valley, while TRD, in Traverse County, is near the South Dakota border on the western plains), exhibit unilateral micromely: all bones of the affected limbs appear to be present. Both animals exhibit a pelvic shift favoring caudal thrust in the "good" leg. Note also that in both animals the ilium on the affected side is also malformed, with heavier ossification near the acetabulum.

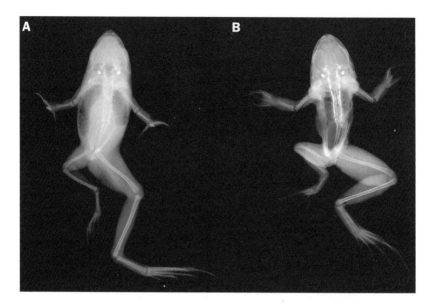

FIGURE 2.24   (A) *Rana pipiens.* 40 mm SUL. Collected on 19 September 1997 from the NEY site by Minnesota Pollution Control Agency field biologists; (B) *Rana pipiens.* 37 mm SUL. Collected by Dave Hoppe at the TRD site in Minnesota during the summer of 2001 and sent to Dan Sutherland for parasite analysis; Dan did not find *Ribeiroia* metacercariae. Frogs shown in Figures 2.28, 2.29A, 2.31, 2.36A, 2.37, 2.40, and 2.55 are also from the NEY site; frogs shown in Figures 2.36B and 2.45 are also from the TRD site. See Chapter 3 for descriptions and Figures 3.2 and 3.4 for photographs of these sites.

HEMIMELIA

Example 25   This northern leopard frog (Fig. 2.25) exhibits a foreshortened limb with a small foot; the long bones of this limb exhibit taumelia (bony triangles). Meteyer and her colleagues[2] call this malformation type phocomelia (p. 161), described as "an abnormal foot attached to a short limb composed of small, disorganized, and unrecognizable bones." The use of the term phocomelia immediately makes one think of the drug thalidomide, which may or may not be a fair association.

Note also the opposite limb, which exhibits a foreshortening of the femur with a focal thickening. It is not possible to determine whether this abnormality is a developmental malformation or a healed broken bone.

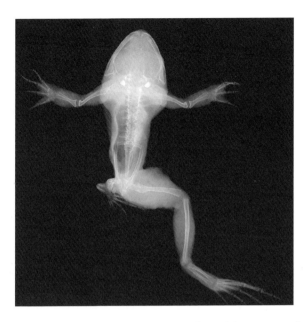

FIGURE 2.25   *Rana pipiens.* 32 mm SUL. Collected by Dan Sutherland in 1998 from a farm pond in Trempealeau County, Wisconsin. According to Sutherland, these frogs did not have metacercariae of the trematode *Ribeiroia ondatrae.* Frogs shown in Figures 2.5, 2.7, 2.11, 2.13, 2.19, 2.20, 2.22A, B, C, and 3.13 represent additional animals collected at Sutherland's Wisconsin site.

Taumelias (bony triangles) are an interesting malformation type. As described by Dave Gardiner and Dave Hoppe,[11] bony triangles occur when a given limb segment is (p. 211) ". . . bent back upon itself, such that the proximal and distal ends of the bone are adjacent to one another and the midpoint of the shaft forms the apex of a triangle . . . The degree of angulation varies from bowing to complete folding." This definition varies from the one given by Meteyer and her collegues,[2] where to be a taumelia, a long bone must be bent back on itself forming an angle > 90°. Gardiner and Hoppe also note that despite being called "bony," these dysplasias are variably ossified/chondrified.

UNILATERAL HEMIMELY

Example 26   Hemimely (foreshortening) of the tibiofibula with an associated taumely. It is unclear whether this abnormal tibiofibula (Fig. 2.26)

FIGURE 2.26   *Rana pipiens*. 32 mm SUL. Collected on 27 July 1998 at the CWB site in Crow Wing County, Minnesota, by Minnesota Pollution Control Agency field biologists. Frogs shown in Figures 2.10, 2.30, 2.33, 2.36D, 2.38, 2.39, 2.41–2.44 and 2.46 are also from CWB. See Chapter 3 for a description and Figure 3.5 for a photograph of this site.

was duplicated or whether the tibia and fibula did not fuse and are thickened. Distal to this region, the tibiofibulare and foot bones appear roughly normal in size and proportion.

BILATERAL ECTROMELY

Example 27   This northern leopard frog (Fig. 2.27) exhibits bilateral foot and ankle malformations including bilateral ectromely (missing limb segments) with hemimely (foreshortening) of the feet and unilateral taumely of the tibiofibulare. This malformation is a mirror image of the one shown in Meteyer and colleagues'[2] figure 4e.

Animals such as this form one of the battlegrounds for arguments about causes of malformations. Parasitic cysts are known to mechanically disrupt apical ridge ectoderm (AER; see Chapters 1, 5, and 6)[14] and to physically disrupt other tissues, altering the polarity of developing limbs.[15]

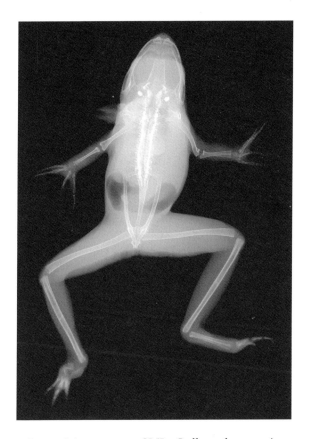

FIGURE 2.27 *Rana pipiens.* 42 mm SUL. Collected on 17 August 1997 from the BLO site in Ottertail County by Minnesota Pollution Control Agency field biologists. See Chapter 3 for a description and Figure 3.11 for a photograph of this site.

Given that limbs form from proximal to distal, and given that Johnson and his colleagues[16] observe that *Ribeiroia ondatrae* ". . .is the only trematode known to cause amphibian malformations" and that[17] ". . .metacercariae [are] highly localized in the tissue around the pelvic girdle and hindlimb, often in close association with abnormal or extra limbs," developmental biologists have wondered how metacercarial cysts can cause such distal malformations. The question, stated more succinctly, is: How can cysts associated with the pelvis alter development in the foot without affecting intermediate structures? This question is addressed in Chapter 5.

Examples 28, 29   Animals presenting with an immobile, straight leg constitute a surprisingly common malformation type. The key presentation is a knee locked in extension. The hip joint is also immobile and radiographs show either an abnormal pelvis (Fig. 2.28) or a disarticulation between the femur and the pelvic girdle (Fig. 2.29A). We do not know whether this disarticulation is due to primary developmental effects, secondary to having extended limb joints, or represents some combination

FIGURE 2.28   *Rana pipiens.* 36 mm SUL. Collected 29 September 1997 at the NEY site in LeSueur County, Minnesota, by Minnesota Pollution Control Agency field biologists. Animals shown in Figures 2.24A, 2.29A, 2.31, 2.36A, 2.37, 2.40, and 2.55 are also from the NEY site. See Chapter 3 for a description and Figure 3.2 for a photograph of this site.

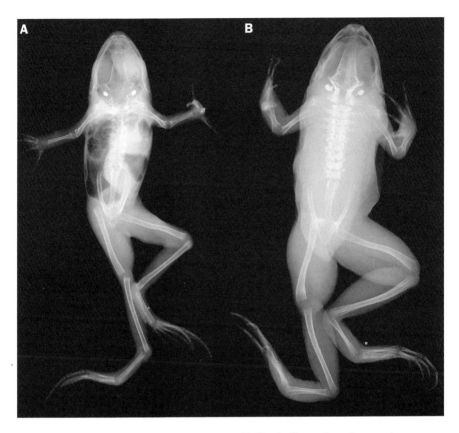

FIGURE 2.29 (A) *Rana pipiens.* 38 mm SUL. Collected 19 September 1997 at the NEY site in LeSueur County by Minnesota Pollution Control Agency field biologists; (B) *Rana catesbeiana.* 50 mm SUL. Collected in California during the 1997–1998 field season by Pieter Johnson. Frogs shown in Figures 2.24A, 2.28, 2.31, 2.36A, 2.37, 2.40, and 2.55 are also from the NEY site, described in Chapter 3 and shown in Figure 3.2. See Figures 2.3, 2.32, 2.35, 2.51 and 3.15 for images of other animals from Pieter Johnson's site.

of primary and secondary effects. The ankle joint of every animal expressing this malformation type is also in extension (confusingly called dorsiflexion [as opposed to plantarflexion, which is true flexion]), as are foot and toe joints. This malformation type has been termed "hyperextension," a term that should not be confused with the condition in human sports injuries. With the exception of the disarticulated hip joint, which in some animals may be due to secondary effects, no joint in these animals is literally hyperextended, although all joints in the affected limb are in extension. Perhaps a better term would be "polyextension."

While limbs affected by hyperextension present a characteristic appearance, the condition of the pelvis is variable. For example, in Figure 2.29, the pelvis of both animals is intact, the pelvic bones are proportional and the correct size, and the affected femur appears articulated. Comparing the pelvic structure in Figure 2.28 to Figure 2.29A and B, we can see (although densities [perhaps sand] in the large intestine obscure it) a pelvis that is misplaced, shifted not only towards the abnormal side but also displaced rostrally (towards the head end of the animal). The ilium on the abnormal side is short and irregularly shaped, whereas the ilium on the opposite side is about twice as thick as expected, curved, and does not appear to articulate with the sacrum. Measurements suggest that long bones are shorter in straight limbs.

The animals in Figure 2.29 were chosen to show that hindlimb hyperextension is a malformation exhibited across species (*Rana pipiens* and *R. catesbeiana*) and distant sites (Minnesota and California). Forelimb hyperextension is rare, perhaps because of the physical constraints that developing limbs experience in the tadpole gill chamber (see Fig. 2.31).

SKIN WEBBING

Example 30   Developing limbs achieve their proper length and proportion by a combination of addition and selective subtraction (described in Chapter 1). Skin webbing is a common malformation that likely represents the subtraction process gone awry, although Gardiner and Hoppe[11] note (p. 211): "At this point we do not know how skin webbings develop, and thus do not understand the developmental mechanisms involved. We also do not know if there is a functional relationship between abnormal skin development and abnormal skeletal development . . ."

This mink frog (Fig. 2.30) from the CWB site in central Minnesota demonstrates the failure of the skin at the knee, and perhaps the ankle and hip joints, to separate. In addition, and perhaps secondary to this skin webbing, the long bones—femur, tibiofibula, and tibiofibulare—of the affected limb are foreshortened (hemimelic). Pelvic elements appear to have shifted, again to favor forward thrust by the normal limb, producing an unusual juxtaposition of the femoral heads.

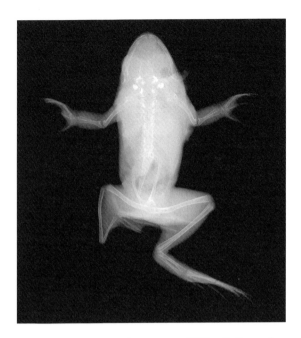

FIGURE 2.30   *Rana septentrionalis.* 44 mm SUL. Collected on 2 July 1998 from the CWB site in Crow Wing County, Minnesota, by Minnesota Pollution Control Agency field biologists. Frogs shown in Figures 2.10, 2.26, 2.33, 2.36D, 2.38, 2.39, 2.41–2.44 and 2.46 are also from CWB. See Chapter 3 for a description and Figure 3.5 for a photograph of this site.

Skin webbing is associated with animals that exhibit a variety of malformation types including bony triangles (taumelia), polymelia, crooked limbs, and pigment abnormalities. Skin webbing is typically not associated with missing limb elements nor with malformations in other body regions such as jaws or eyes. Skin webbing is found only rarely in forelimbs, which is surprising in a sense, given that forelimbs develop in the confined space of the gill chambers.

OTHER FORELIMB PROBLEMS

UNILATERAL HEMIMELIA

Example 31   This northern leopard frog (Fig. 2.31) exhibits a mildly foreshortened humerus and a severely foreshortened radioulna in a forelimb that never emerged from the branchial cavity at metamorphosis. We

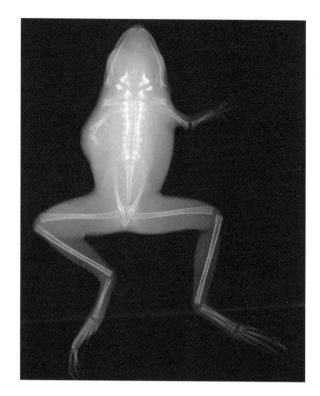

FIGURE 2.31 *Rana pipiens.* 35 mm SUL. Collected on 3 October 1998 from the NEY site in LeSueur County, Minnesota, by Minnesota Pollution Control Agency field biologists. Frogs shown in Figures 2.24A, 2.28, 2.29A, 2.36A, 2.37, 2.40, and 2.55 are also from the NEY site. See Chapter 3 for a description and Figure 3.2 for a photograph of this site.

cannot know whether the bones are foreshortened because of failure to emerge (and the constraints of developing in a confined space), or whether the limb failed to emerge because the long bones were short. This malformation type is rarely observed.

OTHER PROBLEMS, BOTH LIMBS

ARTHROGRYPOSIS

Example 32 A population of bullfrogs from California (where they are not native) exhibits an unusual form of malformation (Fig. 2.32). In these animals, each of their long bones, as well as some other bones, exhibits a kink. Kinks occur at the position where we expect perforating arteries to penetrate. Affected bones include all limb bones—fore- and hindlimbs, bi-laterally—and the iliums. Bones not affected include the coccyx, verte-brae, scapulae, coracoids, and those composing the skull. The mandibles (which form the lower jaw) have no kinks, but are not fully formed (the forward portions remain cartilage). In addition to the skeletal problems, the tails of these animals are not fully resorbed.

All animals from this population, which were contributed by Pieter Johnson,[3] exhibit these malformations; we have not seen a malformation

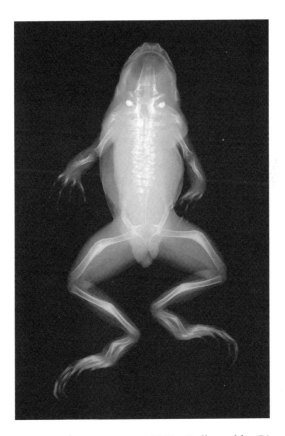

FIGURE 2.32 *Rana catesbeiana.* 44 mm SUL. Collected by Pieter Johnson from a site in California, where bullfrogs are non-native. According to Johnson, these animals were infected by *Ribeiroia ondatrae* metacercariae. See Figures 2.3, 2.29B, 2.35, 2.51 and 3.15 for images of other animals from this site.

pattern like this from any other bullfrog population or from populations of any other amphibian species. These animals were infected by *Ribeiroia ondatrae* metacercariae. Given this, it is interesting that none of the malformations occur in the inguinal region, where cysts of this parasite are commonly found; instead these malformations occur systemically.

While the accepted term for "bent bones" is arthrogryposis, this morphology actually falls within the definition of bony triangles (or taumelia) as defined by Gardiner and Hoppe[11] when they note (p. 211): "The degree of angulation varies from bowing to complete folding to form a triangle."

FIGURE 2.33 *Rana pipiens.* 34 mm SUL. Collected on 19 August 1998 from the CWB site in Crow Wing County by Minnesota Pollution Control Agency field biologists. Arrows indicate positions of perforating nutrient arteries. Frogs shown in Figures 2.10, 2.26, 2.30, 2.36D, 2.38, 2.39, 2.41–2.44 and 2.46 are also from CWB. See Chapter 3 for a description and Figure 3.5 for a photograph of this site.

Example 33 This northern leopard frog (Fig. 2.33) exhibits bilateral "bent" tibiofibulas and a unilateral "bent" radioulna. Note that in the tibiofibulas the bends are symmetrical and occur immediately proximal to the positions of the perforating arteries.

TOO MANY HINDLIMBS: POLYMELY

Example 34 Polymelic animals serve as the iconic image of malformed frogs, but compared to ectromelic animals, they are rare. What is so different, and therefore so striking, about this polymelic mink frog (Fig. 2.34) is that the extra limb appears at a point that is anatomically inappropriate. It is as if the femur has two knees: a normal one that forms a joint that is

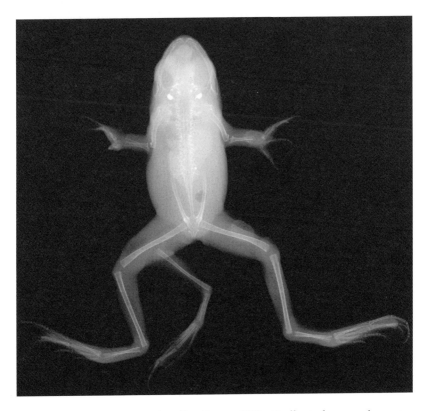

FIGURE 2.34 *Rana septentrionalis.* 38 mm SUL. Collected on 19 August 1997 from the NLA site by Minnesota Pollution Control Agency field biologists.

mobile and acts normally, and a second one about halfway down its length that does not appear to be a joint at all, but supports a tibiofibula and its attached foot. The normal limb is the same size as the contralateral limb, and in life worked about as well; the second limb is less robust and exhibits a shorter tibiofibula and smaller foot.

Example 35 Some limbs seem determined to divide. This Pacific treefrog (Fig. 2.35) collected by Pieter Johnson in 1997–1998 is normal except for a patterned duplication of hindlimb segments. Grossly, this animal was described accurately as having a thick thigh, a duplication at the knee, and a duplication at one ankle producing a small foot. The radiograph

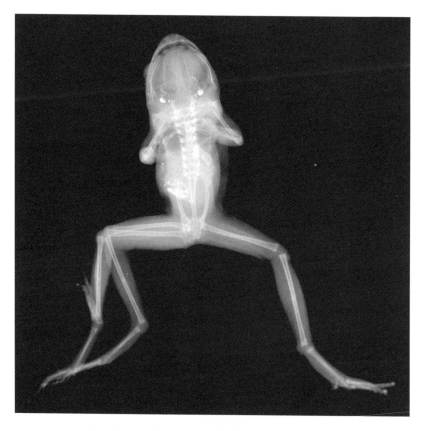

FIGURE 2.35 *Pseudacris regilla.* 20 mm SUL. Collected from a California wetland by Pieter Johnson during the 1997–1998 field season. See Figures 2.3, 2.29B, 2.32, 2.51 and 3.15 for images of other animals from this site.

shows that the duplication actually occurs at the hip—that the thick thigh is the result of two femurs that share a knee joint. The pelvis appears to support two acetabulums, one for each femur, with the assist of a thickened ilium. The femur that articulates at the level of the contralateral femur appears to be the least normal—it is short, kinked, and appears to join the lower limb segment that divides at the ankle. The second femur articulates immediately caudally (towards the rear); it is about 8% shorter than the contralateral femur, but supports lower limb segments comparable in length and girth to those on the normal side.

As unusual as this malformation type would seem to be, northern leopard frogs collected from several Minnesota hotspots, as well as a wood frog from an Alaskan Wildlife Refuge (Fig. 2.36C), exhibit a similar pattern. According to Meteyer et al.[2] (p. 157), polymelia was found in frogs collected at two Minnesota sites during 1997 and 1998 and represented 5.4% (31 of 570) of all the malformations documented during field surveys at Minnesota sites.

Example 36   This northern leopard frog (Fig. 2.36A) exhibits a unilateral polymely, with the duplicated limb segment smaller than the primary limb segment. A similar 32 mm SUL leopard frog was collected on 19 August 1998 from CWB by Minnesota Pollution Control Agency field biologists. Note that while the duplication appears to the naked eye to occur at the knee joint, radiography reveals that the duplication actually occurs at the hip; the duplicated femurs are not associated with distinct limb segments. A close examination of the hip reveals evidence of two acetabulums (hip joints). The knee consists of a tight joint between the caudal femur and the duplicated limb and a loose joint between the rostral femur and the primary tibiofibula. This northern leopard frog (Fig. 2.36B) from a different Minnesota site exhibits a similar morphology, except that the distal ends of the femurs diverge. A wood frog from the Kenai National Wildlife Refuge in Alaska (Fig. 2.36C). In this animal the femur on the polymelic side is normally sized; the duplication occurs at the knee joint and consists of one normal-sized lower limb and one small lower limb. A northern leopard frog that exhibits unilateral duplication without separation of the tibiofibula and duplication with separation distal to this point (at the ankle joint) involving the tibiofibulares and the feet (Fig. 2.36D). The ankle joint itself is fixed in plantarflexion.

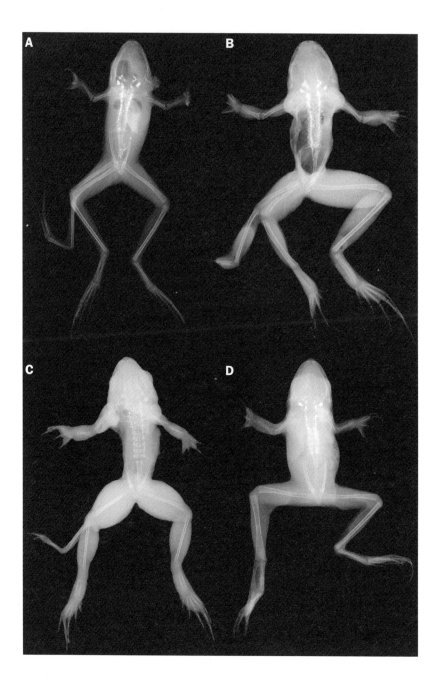

FIGURE 2.36 Opposite. (A) *Rana pipiens*. 37 mm SUL. Collected on 31 August 1998 at the NEY site in LeSueur County, by Minnesota Pollution Control Agency field biologists; (B) *Rana pipiens*. 35 mm SUL. Collected by Dave Hoppe during the summer of 2001 at the TRD site in Traverse County and sent to Dan Sutherland for parasite analysis; (C) *Rana sylvatica*. 22 mm SUL. Collected in 2003 on the Kenai National Wildlife Refuge in Alaska by Mari Reeves;[6] (D) *Rana pipiens*. 35 mm SUL. Collected on 19 August 1998 at the CWB site in Crow Wing County by Minnesota Pollution Control Agency field biologists. Dan Sutherland reported no *Ribeiroia ondatrae* metacercariae in animals collected at either the TRD site (B) or in Alaska (C); the NEY and CWB sites in Minnesota are both known to harbor high densities of *Ribeiroia*. Frogs shown in Figures 2.24A, 2.28, 2.29A, 2.31, 2.37, 2.40, and 2.55 are also from the NEY site. Frogs shown in Figures 2.24B and 2.45 are also from the TRD site. Frogs shown in Figures 2.10, 2.26, 2.30, 2.33, 2.38, 2.39, 2.41–2.44 and 2.46 are also from the CWB site. See Chapter 3 for descriptions and Figures 3.2, 3.4, and 3.5 for photographs of these wetlands.

The animals here illustrate the difficulty in determining the cause of malformations from malformation types. Animals shown in (A) and (D) were collected at sites known to harbor the trematode parasite *Ribeiroia ondatrae*; sites (B) and (C) have been well sampled and are not known to harbor *Ribeiroia ondatrae* (in fact, Dan Sutherland never found *Ribeiroia ondatrae* in any animal collected from Alaska). The animals here and the animal in Figure 2.35 also illustrate, once again, that the same malformation type can be found in different species (even different families) and at widely scattered sites (Minnesota, California, Alaska).

Example 37   The northern leopard frogs in Figures 2.26 and 2.27 represent cases of hemimely (shortened limb elements) associated with, and likely caused by, taumely (bony triangles). The northern leopard frog shown here (Fig. 2.37) also exhibits hemimely with taumely, but additionally demonstrates limb multiplication; there are three hindlimbs in this complex in addition to two bones near the pelvis—at least one of these is taumelic—which may represent additional femurs.

Taumely may or may not be associated with duplicated limb elements, something that Gardiner and Hoppe[11] note only tangentially (p. 211):

FIGURE 2.37 *Rana pipiens.* 41 mm SUL. Collected on 19 September 1997 at the NEY site in LeSueur County, Minnesota, by Minnesota Pollution Control Agency field biologists. Frogs shown in Figures 2.24A, 2.28, 2.29A, 2.31, 2.36A, 2.40, and 2.55 are also from the NEY site. See Chapter 3 for a description and Figure 3.2 for a photograph of this site.

"We have observed bony triangles at all levels along the proximal–distal axis of both primary and secondary limbs at equal frequency in each."

Example 38   This northern leopard frog (Fig. 2.38) exhibits a unilateral polymely, with two complete feet present on the affected side. In this regard this animal resembles the animals illustrated in Figures 2.35 and 2.36.

FIGURE 2.38   *Rana pipiens.* 35 mm SUL. Collected on 27 July 1998 at the CWB site by Minnesota Pollution Control Agency field biologists. This skeleton is lightly ossified. Frogs shown in Figures 2.10, 2.26, 2.30, 2.33, 2.36D, 2.39, 2.41–2.44 and 2.46 are from CWB. See Chapter 3 for a description and Figure 3.5 for a photograph of this site.

In addition, note the asymmetrical pelvis and the shortened, bent femur on the "normal" limb of the affected side. Two extra bony elements are also present, one associated with the proximal (pelvic) end of the femur and one associated with the distal (knee) end.

Example 39   This northern leopard frog (Fig. 2.39) exhibits an unusual form of polymely. Note that the secondary limb is smaller and begins with a displaced portion of the proximal tibiofibula, which does not appear to contact the femur. The two limbs share an ankle joint. Distal to the ankle joint the tibiofibulares are separated but the limb segment itself is not divided. Although obscured in this radiograph, the two feet are the same size, but smaller than the contralateral "normal" foot.

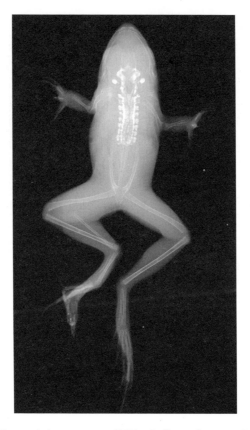

FIGURE 2.39   *Rana pipiens.* 35 mm SUL. Collected on 19 August 1998 at the CWB site in central Minnesota by Minnesota Pollution Control Agency field biologists. Frogs shown in Figures 2.10, 2.26, 2.30, 2.33, 2.36D, 2.38, 2.41–2.44 and 2.46 are from CWB. See Chapter 3 for a description and Figure 3.5 for a photograph of this site.

Example 40   This animal exhibits unilateral taumely associated with the multiplication of distal limb elements (Fig. 2.40). Note that when viewed grossly, the triplication appears to begin at the knee joint; radiography reveals that the femur exhibits a duplication without separation. Note also that the taumely affects one of the tibiofibulas extremely (deep acute angle), one less so (right angle), and one not at all. In this context it is interesting to consider gradients of effect.

FIGURE 2.40 *Rana pipiens.* 39 mm SUL. Collected on 19 September 1997 at the NEY site in LeSueur County by Minnesota Pollution Control Agency field biologists. Frogs shown in Figures 2.24A, 2.28, 2.29A, 2.31, 2.36A, 2.37, and 2.55 are also from the NEY site. See Chapter 3 for a description and Figure 3.2 for a photograph of this site.

Example 41 This northern leopard frog (Fig. 2.41) exhibits a right tibiofibula anteversion (corkscrew-like morphology) accompanied by taumelia (as it almost always is), which splits to form two incompletely formed ankles, one relatively normal foot, and one ectromelic/hemimelic foot.

Example 42 This northern leopard frog (Fig. 2.42) exhibits a left tibiofibula taumely with either extra digits or a partially duplicated foot. Another long bone is present that originates at the knee joint and extends distally. A kink is present about two-thirds of the way along the length of this bone.

Example 43 Typically, the more complicated the malformation, the more complicated the description (e.g., Fig. 2.47). But this example of a polymelic frog (Fig. 2.43) can be portrayed fairly simply. This animal

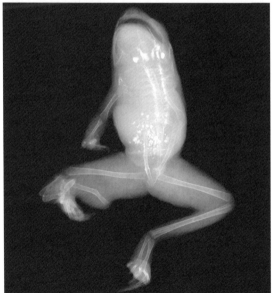

exhibits four hindlimb fields that arise on one side of the pelvis. These limb fields produce one normal-sized hindlimb and three smaller hindlimbs. The smaller hindlimbs are about equally sized. On both sides of the pelvis, the iliums appear thickened and abnormally shaped. There may be extra ilial elements present.

FIGURE 2.41 Opposite. *Rana pipiens.* 30 mm SUL. Collected by Dave Hoppe in 1999 at the CWB site in central Minnesota (specimen # CWB9187024L). Skeleton lightly ossified. Frogs shown in Figures 2.10, 2.26, 2.30, 2.33, 2.36D, 2.38, 2.39, 2.42–2.44 and 2.46 are from CWB. See Chapter 3 for a description and Figure 3.5 for a photograph of this site.

FIGURE 2.42 Opposite. *Rana pipiens.* 33 mm SUL. Collected by Dave Hoppe in 1999 at the CWB site in central Minnesota (specimen # CWB9187083L). Frogs shown in Figures 2.10, 2.26, 2.30, 2.33, 2.36D, 2.38, 2.39, 2.41, 2.43, 2.44 and 2.46 are from CWB. See Chapter 3 for a description and Figure 3.5 for a photograph of this site.

FIGURE 2.43 Above. *Rana septentrionalis.* 40 mm SUL. Collected by Dave Hoppe at the CWB site in central Minnesota (specimen # CWB7168012M). Dave Hoppe has taken good photos of this animal both alive and as a cleared and stained specimen. Frogs shown in Figures 2.10, 2.26, 2.30, 2.33, 2.36D, 2.38, 2.39, 2.41, 2.42–2.44 and 2.46 are from CWB. See Chapter 3 for a description and Figure 3.5 for a photograph of this site.

It would be good to examine the (likely diminished) nature of locomotory performance arising from the presence of multiple smaller hindlimbs and ask whether it affects the survival of these animals.

According to the survey of Meteyer and her colleagues[2] (p. 155), supernumerary limbs that originated from the pelvis or from bones resembling duplicate pelvic elements were found in seven out of 570 frogs.

Example 44   A mid-metamorphic mink frog tadpole with right hindlimb polymelia (Fig. 2.44). This animal was cage reared at the CWB site, and Dave Hoppe reports that the extra limbs appeared well after primary limb development. While no one would think to call a morphology such as this cancerous, conceptually this is not the stretch it would seem to be. Stedman's Medical Dictionary defines cancer as a neoplasm, and neo-

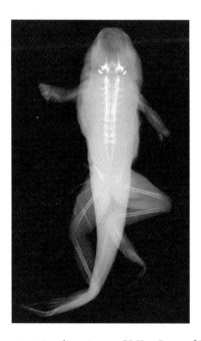

FIGURE 2.44   *Rana septentrionalis.* 38 mm SUL. One of Dave Hoppe's experimental animals from the CWB site in Crow Wing County, Minnesota (specimen #CWB6258002M). Frogs shown in Figures 2.10, 2.26, 2.30, 2.33, 2.36D, 2.38, 2.39, 2.41–2.43 and 2.46 are also from CWB. See Chapter 3 for a description and Figure 3.5 for a photograph of this site.

plasm literally means "new growth." This is an animal that kept growing hindlimbs after a period when hindlimb production should have stopped.

Example 45   The pelvis of this northern leopard frog (Fig. 2.45) exhibits a slight shift favoring the positioning of the good limb, but the pelvic bones are normal. There are three femurs emerging from the affected side; none of them are normal. One is truncated (ectromelic) and two give rise to micromelic limbs. Both femurs in the micromelic limbs are bent— more misshapen than the other bones in these limbs. The largest of the micromelic limbs originates from behind the pelvic girdle.

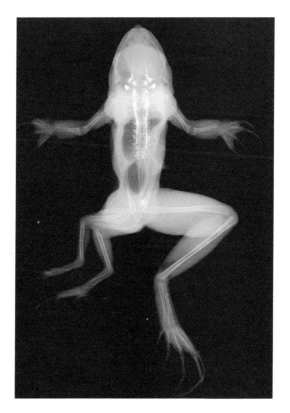

FIGURE 2.45   *Rana pipiens.* 41 mm SUL. Collected by Dave Hoppe at the TRD site in Traverse County, Minnesota, during the summer of 2001 and sent to Dan Sutherland for parasite analysis. Frogs shown in Figures 2.24B and 2.36B are also from the TRD site. See Chapter 3 and Figure 3.4 for a description and a photograph of this site.

Example 46   This northern leopard frog (Fig. 2.46) exhibits an extra left hindlimb that appears to fuse medially at the groin. Dave Hoppe reports it was recorded in the field as splitting at the ankle because both femurs and tibiofibulas were located in one muscle mass. The pelvis is shifted toward the affected side.

Less noticeable (since we're looking at the pelvis and hindlimbs) is the abnormal shape of the skull, which is comparatively narrow and pointed for a leopard frog. It is important to remember that malformations can come from localized causes (which can be multiple) or from systemic-level causes that can selectively affect different regions of the body or different organ systems (think about chromosomal abnormalities and the suites of effects they can produce). Oftentimes we equate importance with impact, focusing on the obvious and ignoring the subtle, when in fact the subtle can provide the more important clue.

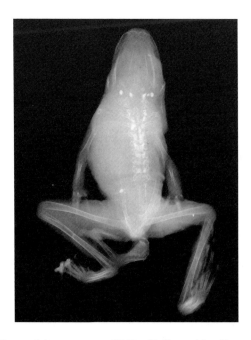

FIGURE 2.46   *Rana pipiens.* 31 mm SUL. Collected by Dave Hoppe at the CWB site in Crow Wing County, in central Minnesota (specimen # CWB8197028L). Frogs shown in Figures 2.10, 2.26, 2.30, 2.33, 2.36D, 2.38, 2.39, and 2.41–2.44 are also from CWB. See Chapter 3 for a description and Figure 3.5 for a photograph of this site.

Example 47 Some malformations are so complex that the detail is difficult to understand. Harvey Cushing, the father of neurosurgery said:[18] ". . . when something is so colossal as to transcend comprehension one must reduce it to the simple terms of familiar things." This process is, in fact, not a bad description of how best to do science.

The most difficult malformations to describe are those with multiple segments that present in different ways. The bullfrog represented here (Fig. 2.47) is a museum specimen housed in the Hefner Zoology Museum at Miami University in Oxford, Ohio. It was collected in 1954–1955 (MU AR163, AR648–AR653, MU) by L. Campbell and P. Daniel.[19] This animal and seven others were collected from a site near Ripley, Ohio. The other animals are much smaller, and likely younger, but all exhibit the same types of malformations (see Fig. 4.14). The historic malformations discovered in this population were detailed in an unpublished Master's thesis by Hauver[20] and re-described in an interesting paper by Pieter

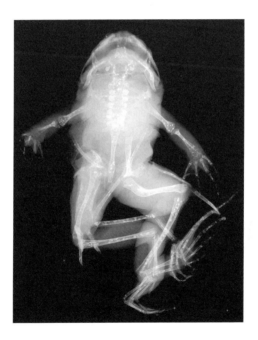

FIGURE 2.47 *Rana catesbeiana.* 56 mm SUL. Specimen collected from a wetland near Ripley, Ohio. According to Johnson and his colleagues, 19 of these animals contained no *Ribeiroia ondatrae* metacercariae. Additional animals from the Ripley site are shown in Figure 3.14.

Johnson and his colleagues.[19] This Ripley Pond site is featured in Chapter 3, and the importance of these animals is discussed in Chapters 5 and 7.

Aside from the abnormal calcifications, which were likely produced by the animal rather than as an artifact of preservation and storage, the axial skeleton, jaws, and forelimbs all appear to be normal (although we concede that this is more difficult to confirm in animals fixed in a contorted position than in animals fixed in an anatomical position designed to visualize skeletal elements unambiguously). The problems here arise with the pelvis and hindlimbs and fall into several categories:

Too many elements
Elements out of position
Elements the wrong size
Elements the wrong shape
Elements that cannot be identified

Usually, multilegged animals such as this suggest *Ribeiroia ondatrae* parasites are the cause, but Johnson and his colleagues[19] found no *Ribeiroia* in any animals, historic or recently collected, from this site.

OTHER MALFORMATION TYPES

HYGROMA

Example 48   Some of the most unusual patterns of malformations involve multiple effects. Bullfrog tadpoles collected by Charles Facemire from a population in Switzerland County, Indiana (Fig. 2.48), exhibited bilaterally duplicated hindlimbs in combination with subcutaneous swellings caused by accumulations of serous fluid (original description in our 2003 paper[3]). These swellings extend from joint to joint—hip to knee, knee to ankle, and extend around each foot. Vic Jolgren, a pathologist and former colleague (now retired), suggested calling these hygromas. Four animals from this population have been radiographed (see Fig. 3.16 for additional animals) and all show some form of hygroma formation; two animals, the one shown here and one additional animal (Fig. 3.16A), exhibit multiple hindlimbs. The year following the collection of these

FIGURE 2.48   *Rana catesbeiana.* ~26 mm SUL. Collected by Charles Facemire from a wetland in Switzerland County, Indiana. Additional animals from this site are shown in Figure 3.16.

animals, Facemire returned to collect additional animals, but was disappointed. Not only did he fail to find malformed frogs, he failed to find any frogs. The population had been extirpated.

JAW

Example 49   This northern leopard frog (Fig. 2.49) has an incompletely formed upper jaw (maxilla) and resembles, at least superficially, cleft lip (frogs have no palates) seen in humans. The JOF site was considered a marginal hotspot; no other animals exhibiting this malformation were collected there.

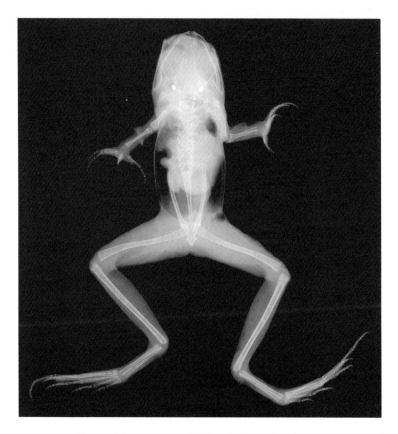

FIGURE 2.49   *Rana pipiens.* 45 mm SUL. Collected 5 August 1997 at the JOF site by Minnesota Pollution Control Agency biologists.

BLOAT

Example 50   This northern leopard frog (Fig. 2.50) exhibits bloating. In general, bloating can be due to either fluid or air build up. In this animal, it appears as though air (less dense) is responsible on the left side of the radiograph and fluid and viscera (more dense) on the right side.

UNRESORBED TAIL

Example 51   This bullfrog (Fig. 2.51) is from the same California population represented by the animal illustrated in Figure 2.32 and highlighted in Chapter 3 (Fig. 3.15; see also the description in our previous

FIGURE 2.50 *Rana pipiens.* 52 mm SUL. Collected on 11 August 1997 at the LMS site in Meeker County by Minnesota Pollution Control Agency field biologists.

paper[3]). Unresorbed tails are common, may or may not be associated with other malformations, and appear to have many causes. Among amphibians, unresorbed tails are a problem only for frogs and toads; salamanders retain their tails through, and following, metamorphosis.

Example 52   The tail in this Blanchard's cricket frog (Fig. 2.52) is both shriveled and calcified. The Old Barn site in Indiana where this animal was collected is a tree nursery.[21] According to the landowner, a variety of herbicides have been, and continue to be, used at this site. Sonar, (fluridone: 1-methyl-3-phenyl-5-[3-(triphluorolmethyl) phenyl]-4(1H)-pyridinone)

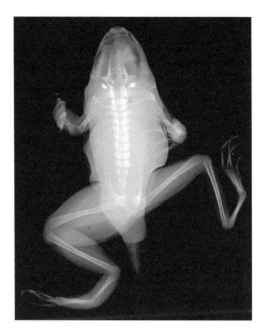

FIGURE 2.51   *Rana catesbeiana.* 50 mm SUL. Collected by Pieter Johnson from a site in California during the 1997–1998 field season. See Figures 2.3, 2.29B, 2.32, 2.35 and 3.15 for images of other animals from this site.

is applied around wetlands about every two years. As with the state-run aquaculture sites in Iowa, Aquazine, (simazine: 2-chloro-4,6-bis (ethylamino)-s-triazine) is being used two or three times a year to control aquatic vegetation. Other chemicals used on the property (> 10 m from the wetland) include the brand names Oust, (sulfometuronmethyl: methyl 2-[[[[(4,6-dimethyl-2-pyrimidinyl) amino]-carbonyl] amino] sulfonyl] benzoate), Stinger, (clopyralid: 3,6-dichloro-2-pyridinecarboxylic acid, monoethanolamine salt), and Pendulum, (pendimethalin: N-(1-ethylpropyl)-3,4-dimethyl-2, 6-dinitrobenzenamine).

TRUE SCOLIOSIS

Example 53   Like the secondary scoliosis produced in amelic animals as a functional response to having only one hindlimb, true scoliosis occurs

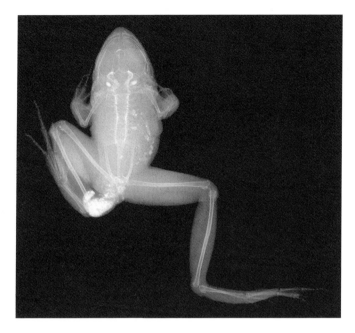

FIGURE 2.52 *Acris crepitans.* 21 mm SUL. Collected by Laura Blackburn on 15 September 1999 at her "old barn" site in east-central Indiana.

relatively frequently (Fig. 2.53). It is most often seen in tadpoles, which have a much longer vertebral column than adults. In tadpoles, true scoliosis is often seen in the proximal portion of the tail and disappears as the tail is resorbed during metamorphosis. In this newly metamorphosed northern leopard frog, scoliosis occurs at about the mid-abdominal level and appears to leverage the proximal portion of the urostyle.

PIGMENT

Example 54   Pigment disruptions on the proximal portions of limbs with missing or malformed distal segments offer powerful proof that these malformations are not due to failed predation attempts. It is difficult to imagine trauma inducing pigment disruptions away from the site of the alleged attack, especially in the absence of scarring. Over 70% of the ectromelic animals collected by the Minnesota Pollution Control Agency during the

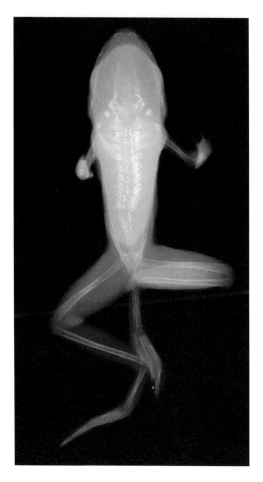

FIGURE 2.53   *Rana pipiens.* 32 mm SUL. Collected on 7 August 1997 at the WIN site by Minnesota Pollution Control Agency biologists. The animals shown in Figures 2.2, 2.12, and 2.23 are also from this site.

1997 and 1998 field seasons exhibited pigment disruptions. These disruptions on the thighs of ectromelic northern leopard frogs included the absence of spotting, spots of different size (usually smaller) when compared to the contralateral limb and to animals from the same population, and spots of a different orientation (orthogonally oriented spots in the example shown in the color photo in Color Plate 4).

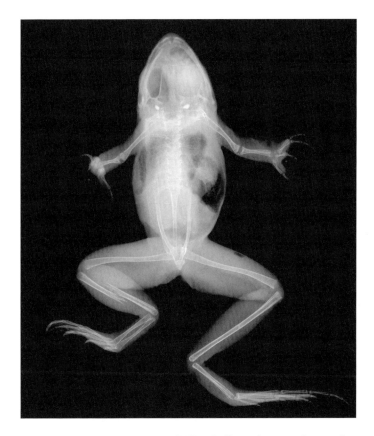

FIGURE 2.54 *Rana pipiens.* 28 mm SUL. Collected on 24 September 1997 at the SUN site in Ottertail County by Minnesota Pollution Control Agency field biologists. Animals from this site are also shown in Figures 2.16 and Color Plate 4.

TRAUMA THAT RESEMBLES MALFORMATION

Example 55   This northern leopard frog (Fig. 2.54) exhibits a compound fracture of the tibiofibula, due to trauma.

Example 56   This animal (Fig. 2.55) exhibits bilateral fractured tibiofibulas. These breaks appear to be caused by trauma, as if from a collector's dip net or another type of accident. These fractures may have been healing at the time of euthanization.

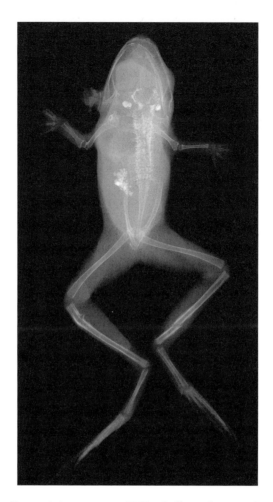

FIGURE 2.55   *Rana pipiens.* 32 mm SUL. Collected on 31 August 1998 from the NEY site in LeSueur County by Minnesota Pollution Control Agency field biologists. Frogs shown in Figures 2.24A, 2.28, 2.29A, 2.31, 2.36A, 2.37, and 2.40 are also from the NEY site. See Chapter 3 for a description and Figure 3.2 for a photograph of this site.

THREE

HOTSPOTS

If you don't know the ground you are probably wrong
about nearly everything else.

NORMAN MACLEAN[1]

In amphibian conservation biology, the term "hotspot" is used in two
contrasting ways. One way, with a good implication, is to denote a site
or a region with high amphibian richness. These places are usually located in tropical or subtropical ecosystems and are often the focus of intense conservation efforts. The second use of the term "hotspot," and the
way I use it here, is much less desirable and indicates a site with a large
number or percentage of malformed animals. The Minnesota Pollution
Control Agency, for example, defined a wetland as a hotspot if, at any
time when it was sampled, it exhibited five percent or more animals with
malformations.

In designating places as malformed frog hotspots, it is important to understand that this can be an arbitrary designation (see "The Nature of the
Malformed Frog Data" in Chapter 5). Further, there are likely many more
hotspots in existence than have been identified. And finally, it is important to know that some malformed frog hotspots wink on and off; specific sites can be hotspots during some years but not others. Understanding
the nature of these hotspots is instructive, because when terms such as
"grotesque" and "monstrosity" are used to describe malformed frogs, we
envision toxic waste dumps, and what comes to mind is a wetland ooz-
ing bubbling green gunk and dead fish with Xs for eyes scattered along
the shore. In reality, most hotspots appear at first glance to be perfectly
normal wetlands, given their placement on a highly altered landscape.

In this chapter I describe some of the best-known hotspot wetlands from Minnesota; I also show a Minnesota control site. These Minnesota sites were considered from an ecoregion perspective.[2] I illustrate a state-run aquacultural site from northwestern Iowa that produces malformed frogs. To make comparisons, I show four sets of radiographs of animals from malformation sites in other parts of the country, including: (1) northern leopard frogs from Dan Sutherland's Trempealeau County, Wisconsin, site; and sets of bullfrog radiographs from (2) the Ripley Pond site in Ohio,[3] (3) a Santa Clara County, California, site identified by Pieter Johnson,[4] and (4) a Switzerland County, Indiana site.[4] Finally, I briefly describe the U.S.F.W.S. National Wildlife Refuge sampling program. Some generalizations about these sites will apply broadly, some will not.

MINNESOTA STUDY SITES

In 2001, the University of Wisconsin parasitologist Dr. Daniel Sutherland[5] and I were asked to visit and re-assess the hottest of the Minnesota malformed frog hotspots. Dan brought along his graduate student, Josh Kapfer, and together we sampled 17 sites (Figure 3.1).[4] The majority (11) of these sites were considered hotspots. Four Minnesota and two Iowa wetlands were used as reference sites.

During the early days of the malformed frog investigation not enough attention was paid to the characteristics of hotspot sites. No wetland ecologists were brought on board to provide descriptions and search for signs of problems. Instead, field teams would arrive, record the time and perhaps the ambient temperature, collect frogs and/or water samples, and leave for the next site.

Our group sampled malformed frogs much in the way that other teams had done; at each site we collected all malformed frogs encountered and a subsample of normal frogs, brought them into the lab, dissected them to determine their parasite loads, then radiographed them. We also collected water samples and shipped them to U.S.G.S. hydrologist Don Rosenberry for analysis. In addition, and beyond what had been done previously, Sutherland, Kapfer, and I described sites in terms of origin (natural, human created, and restored) and malformed frog history (hotspot or reference site; below). We took a close look at each site and considered environmental influences that might be important. For example, we

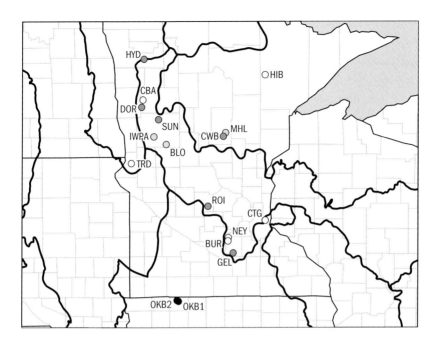

FIGURE 3.1   A map of Minnesota and portions of surrounding states
showing sites that Sutherland, Kapfer, and I sampled during the late
summer of 2001. Dark-gray dots indicate natural wetlands considered
to be hotspots, and are HYD in Polk County, DOR in Becker
County, SUN in Otter Tail County, CWB in Crow Wing County,
ROI in Meeker County, and GEL in LeSueur County. White dots
indicate created wetlands considered to be hotspots, and are CBA in
Becker County, HIB in St. Louis County, TRD in Traverse County,
NEY and BUR in LeSueur County, and CTG in Washington County.
Light-gray dots show Minnesota reference (control) sites associated
with hotspot wetlands, and are IWPA and BLO in Otter Tail County,
and MHL in Crow Wing County. Black dots represent two reference
sites from northwestern Iowa—wetlands long studied by the author and
distant from impacts influencing the Minnesota sites. Bold lines indicate
ecoregion boundaries.[2] Map assembled by Robert Klaver.

found sites in close association with high-end housing developments, low-
end trailer courts, and agricultural fields (at least one site hosted cattle).
Some sites were manmade, most were natural. Some sites were true pot-
holes (with warm standing water), others were spring fed (and cold). Most
sites looked fine at first glance; a couple truly looked skanky, and at these

we played the Alfonse and Gaston routine of "After you," "No, after you" with ourselves. One memorable, and almost personally permanent site looked fine until our first step into the water, when we immediately became stuck in waist-deep muck—the product of severe erosion from the surrounding agricultural fields.

From the 15 Minnesota and two Iowa sites visited, we sampled a total of 837 amphibians. Of these we collected 274 for parasite and radiographic analyses. These animals represented six species (*Bufo americanus*, *Pseudacris triseriata*, *Rana catesbeiana*, *R. clamitans*, *R. pipiens*, and *R. sylvatica*).

The reality is that *most* malformed frog hotspots look fine to the untrained eye, no different from wetlands that produce normal frogs. But impressions are just that, and when you look carefully, several Minnesota sites stand out. For example, the NEY wetland (Fig. 3.2), which started the malformed frog phenomenon, is a large, constructed wetland located in Le Sueur County. It is surrounded by former and active agricultural fields, and is the centerpiece of a recently constructed nature center. In 2001,

FIGURE 3.2   The NEY site as it appeared in August 2001, on the day we sampled. This is a created wetland and an important malformed frog hotspot. Animals shown in Figures 2.24A, 2.28, 2.29A, 2.31, 2.36A, 2.37, 2.40, and 2.55 are from this wetland. A subset of the data shown in Table 5.5 is from this site. Photo by the author.

sago pondweed (*Potamogeton pectinatus*) was the predominant macrophyte, and formed thick beds that restricted water currents and produced pockets of warm surface water. Snails were numerous. Along with Cindy Reinitz, Tim Halliday, and Tony Gamble, we sampled 115 northern leopard frogs.

The ROI site (Fig. 3.3) is a shallow, semipermanent natural wetland in Meeker County, Minnesota. The basin is surrounded by forest on the north and west sides and by mobile homes and lawns on the south and the east sides. More trash was found here than at any site except the park in the St. Paul suburb of Cottage Grove (CTG). Duckweed (*Lemna minor*) covered over 80% of the water's surface. Emergent cattails (*Typha* sp.) and grasses—indicating a history of drying—were present. A large number of invertebrate species were observed. One American toad, 11 leopard frogs, and four wood frogs were sampled. We also captured an eastern gray treefrog tadpole and one salamander in the *Ambystoma laterale* complex. A chemical analysis of the water confirmed what we suspected—that there was septic system "leakage" into the wetland.[4]

FIGURE 3.3   The ROI site as it appeared in August 2001, on the day we sampled. This is a natural wetland and one of the hottest of the Minnesota hotspots. Frogs shown in Figures 2.1 and 2.22D are from this wetland. Photo by Dan Sutherland.

The TRD site (Fig. 3.4) is a small, but deep, constructed wetland on public school grounds located in Traverse County. The basin is composed of hardpan clay with steep sides and is ringed by cattails. Well water is pumped in periodically to prevent pond drying. During our sampling visit, macrophytes had not yet established themselves, although by September, coontail (*Ceratophyllum demersum*) was present (D. Hoppe, personal observation). Few invertebrates were observed, and in three years of sampling, no snails of any species have been collected (D. Hoppe, personal observations). Various species of fish, including bullheads (*Amieurus* sp.) and fathead minnows (*Pimephales promelas*), have been introduced, although no fish were observed or captured in 2001. Two American toads and 14 northern leopard frogs were sampled.

The CWB site (Fig. 3.5) is a small natural lake located near Lake Mille Lacs in Crow Wing County. It is surrounded by mixed deciduous/conif-

FIGURE 3.4   The TRD site as it appeared on the day we sampled in August 2001. The TRD site is a created wetland and considered a hotspot. Animals shown in Figures 2.24B, 2.36B, and 2.45 were collected at this wetland. Photo by Dan Sutherland.

FIGURE 3.5 The CWB site as it appeared during the summer of 2001. The eroded grassy bank on the far side is where cattle come down to water. Note the lack of emergent vegetation in the eroded area. CWB is a natural site, and one of hottest of the Minnesota hotspots. Animals shown in Figures 2.10, 2.26, 2.30, 2.33, 2.36D, 2.38, 2.39, 2.41–2.44, and 2.46 were collected at this wetland. A subset of the data shown in Table 5.6 is from this site. Photo by Dave Hoppe. Used with permission.

erous forest interspersed with agricultural land and roads. One lakefront house is present. Cattle from an adjacent dairy farm use a portion of the wetland for watering and have greatly eroded the shoreline bank adjoining the collection site. Cattails (*Typha* sp.) ring the shoreline, with stands of water lilies (*Nuphar* sp.) and arrowheads (*Sagittaria* sp.) beyond the cattails. In 2001, three-way sedge (*Dulichium arundinaceum*) predominated in the areas disturbed by fallen trees and a dock. Sunfish (*Lepomis* sp.), brook sticklebacks (*Culaea inconstans*), and minnows (esp. *Pimephales* sp.) were present, as were a variety of invertebrates, including planorbid snails. This is the only site visited where we observed dead and dying frogs and fishes floating in the water and washed up on shore. Dave Hoppe has described this phenomenon.[6] Along with Hoppe and Bill Souder, we sampled 88 mink frogs and one green frog.

FIGURE 3.6   The HYD site in northwestern Minnesota as it appeared on the morning we sampled. This wetland is natural and a hotspot. The animal shown in Figure 2.17 was collected from this site. Photo by Dan Sutherland.

The HYD site (Fig. 3.6) is a permanent, natural wetland in Polk County. The basin is ringed with cattails (*Typha* sp.), willows (*Salix* sp.), swamp milkweed (*Asclepias incarnata*), and Canada thistle (*Cirsium arvense*). Agricultural fields extend beyond the ring of emergent vegetation. This site is unusual in that the bottom is covered with 50–70 cm of loose muck, likely representing severe erosion from the adjacent fields. Frogs were not found in association with the wetland but instead were found feeding in a mowed field across a gravel driveway. Sixty-five leopard frogs were sampled here.

The CBA site (Fig 3.7) is a constructed, permanent wetland in Ottertail County. This wetland was made by diking a hillside punctuated with springs and fens. The water was cold (8°C below air temperature)—consistent with being spring fed. Submerged dead trees were scattered along the water's edge. In the water there were thick macrophyte beds consist-

FIGURE 3.7   The CBA site in northwestern Minnesota as it appeared in August 2001, on the day we sampled. This site was a created wetland and considered to be a hotspot. Photo by Dan Sutherland.

ing mostly of coontail (*Ceratophyllum demersum*). Brook sticklebacks (*Culaea inconstans*) and mudminnows (*Umbra limi*) were present. A large number of invertebrates were also present, including planorbid snails. One hundred and twenty-nine leopard frog adults and seven leopard frog tadpoles were sampled.

The BUR site (Fig. 3.8) is a small, permanent wetland in Le Sueur County that was originally chosen as a reference site for the nearby NEY hotspot, but malformed frogs were subsequently found here. The wetland is located about 15 m from an agricultural field, which is separated from the wetland by a gravel road. The site is surrounded by willows (*Salix* sp.); emergent vegetation includes rushes (*Scirpus* sp.) and sedges (*Carex* sp.). The center of the wetland was mostly open water with some duckweed (*Lemna minor*). Macrophytes were sparse. Brook sticklebacks (*Culaea inconstans*) were present, as were a variety of aquatic invertebrates, including a planorbid snail species (*Helosoma trivolvis*). Eighty-six leopard frog adults were sampled.

FIGURE 3.8 The BUR site as it appeared on the day we sampled in August 2001. This wetland is created and was a hotspot. A subset of the data shown in Table 5.5 is from this site. Photo by Dan Sutherland.

The CTG site (Fig. 3.9) is a permanent wetland constructed to control storm water runoff in the city of Cottage Grove, a southeastern suburb of St. Paul, in Washington County. The site is located in Pinetop Park and is surrounded by housing developments. The wetland shoreline is partially earthen and partially paved with asphalt. Garbage, including broken lawn chairs, a television, beverage cans, and bottles (mostly broken), litter the shoreline and the wetland bottom. Rainwater enters this basin from the west. This area has fluctuating water levels and contains stands of cattails. The water is colored pea-soup green with phytoplankton. Planorbid snails and clam shrimp (concostracans) were present. Twenty-six American toads were sampled.

The HIB site (Fig. 3.10) is a small, constructed permanent wetland located outside of Hibbing, in St. Louis County. This wetland was built in the shape of a ring (to facilitate winter riding lawn mower races on the ice) in 1996. The lawn grades to the pond edge: rushes (*Scirpus* sp.), sedges

FIGURE 3.9   The CTG site as it appeared on the afternoon we sampled it in August 2001. This is a created wetland, and at the time of our sampling, a newly discovered hotspot. The data shown in Table 5.4 are from this site. Photo by Dan Sutherland.

(*Carex* sp.), and cattails (*Typha* sp.) are present. In the water, sago pond-weed (*Potamogeton pectinatus*) predominates. Snails were observed but not sampled. One American toad, seven wood frogs, and 11 leopard frogs were collected.

The BLO site (Fig. 3.11) consists of two semipermanent to permanent wetlands located immediately east of Block Lake, in Ottertail County. One wetland is forested and located across a road, near lake level. A second wetland is located in an old-field grassland about 200 m upland, separated from the first wetland by forested hillside. At the time of our sampling, both wetlands were nearly dry, making an assessment of macrophytes difficult. Block Lake leopard frogs have been sampled for many years by Dave Hoppe and Bob McKinnell because of the high frequency of burnsi morphs (see Chapter 1). Forty-four metamorphosed leopard frogs and 11 wood frogs were collected.

FIGURE 3.10   The HIB site in northeastern Minnesota on the crisp August morning we sampled. This is a created wetland and designated a hotspot. Photo by Dan Sutherland.

Malformed frog hotspots are brought to the attention of authorities when there is a congruence of amphibian malformations and humans interested in the outdoors. We found one previously unreported hotspot (9.1% malformation frequency) simply by being curious about a roadside wetland. Therefore, we must view the map in Figure 3.1 as the product of a non-systematic sampling effort. If we believe the map to be generally representative (and there is independent evidence of this; see NARCAM[7] for similar patterns), malformed frog hotspots tend to occur in a broad line from northwest to southeast across Minnesota—that is, few reported hotspots occur in the southwestern grassland and northeastern boreal forested portions of the state.

Hotspots can occur in natural or created wetlands; five hotspots were natural wetlands, six were created wetlands. It is important to realize that none of the 17 sites sampled were natural in any true sense. Each hotspot and most reference sites were associated with agriculture or housing develop-

FIGURE 3.11   The upper wetland of the two BLO sites in northwestern Minnesota as it appeared during the August 2001 day we sampled. Note that this basin is almost dry. This wetland is natural and served as one of our reference sites. The data shown in Table 5.7 are from this site. Photo by Dan Sutherland.

ments. Some of the hotspots, such as ROI, HYD and CWB, appeared to be highly affected by human development and agricultural practices.

AN IOWA AQUACULTURE SITE

In Iowa, several natural wetlands have been condemned by the state in order to raise game fish from newly hatched fry stages to fingerling stages. Fingerlings are then seined and transported to large lakes, where they grow to a catchable and keepable size. Game fish fry are as susceptible to predation as any young animals. To reduce the threat, prior to releasing fry, rotenone is applied to these wetlands to kill resident fishes and salamander larvae—potential predators. Then, before collecting fingerlings, the aquatic herbicide aquazine is applied to kill submergent plants that can clog the big seine nets. Several malformed frogs have been collected from the Welch Lake site (Fig. 3.12).

FIGURE 3.12   Sign warning of application of the aquatic herbicide aquazine at Welch Lake in Dickinson County, Iowa. The sign reads: "Warning, Aquatic Herbicide Applied, Do Not Use Water for Spraying, Irrigation, Livestock, or Human Consumption, IA DNR." Despite only casual collecting, several malformed northern leopard frogs (including animals presented in Figs. 2.4 and 2.8) have been collected at this site. Photo by author.

AN UNUSUAL WISCONSIN SITE

A Trempealeau County, Wisconsin, wetland apparently began producing malformed northern leopard frogs when a large-scale poultry producer bought the land and spread chicken waste as fertilizer on the surrounding agricultural fields. These animals exhibit hindlimb ectromelia (missing limb elements). Radiographs show a peculiar, bubble-like expansion of compact bone that is found in most malformed animals (Fig. 3.13) and sometimes in the normal hindlimb of malformed animals (Fig. 2.13). Dan Sutherland never found *Ribeiroia* metacercariae in animals collected at this site.

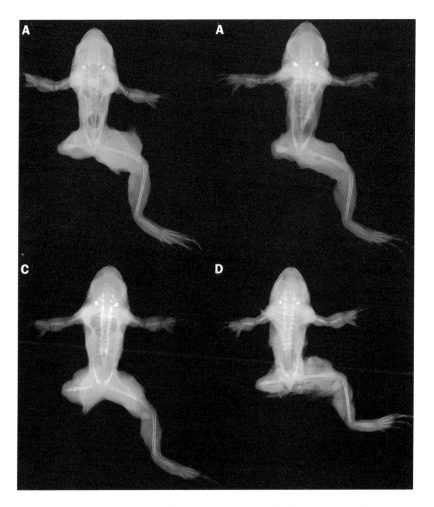

FIGURE 3.13   *Rana pipiens* collected by Dan Sutherland from a farm
pond near Arcadia, Wisconsin, in Trempealeau County in 1998.
Sutherland did not find *Ribeiroia* metacercariae in any of these animals.
All animals shown here demonstrate unilateral hindlimb ectromelia,
although this malformation type does not show up in all animals at this
site. At the point of absence, all animals shown here exhibit a bony expansion that resembles a bubble around the femoral shaft. This "bubble"
morphology is characteristic of most animals found at this site and is
rarely found elsewhere. The animals shown in Figures 2.5, 2.7, 2.11,
2.13, 2.19, 2.20, 2.22A, B, C, and 2.25 were collected from this site.
(A) 33 mm SUL; (B) 31 mm SUL; (C) 32 mm SUL; (D) 32 mm SUL.

Bullfrogs collected during the summers of 1954 and 1955 in a pond near Ripley, Ohio exhibited severe polymelia (Fig. 3.14). Interestingly, animals recently collected at this same site by Pieter Johnson and his colleagues[3] showed no signs of frequent or severe malformations (p. 1731): "One of 25 metamorphosing bullfrogs exhibited a split digit in the left foot . . ." The pond appears to have recovered. Despite searching for *Ribeiroia ondatrae* using both morphological characters and a controversial morphometric identification technique,[8] the authors found no signs of metacercariae in either historic or recent collections, and they suspected chemical or genetic causes of the historic malformations (p. 1726).

Johnson and colleagues[3] write of the malformed animals collected in 1954 and 1955 (p. 1729): "Ripley pond also supported deformed bullfrogs, but the malformations were strikingly diverse and much more severe. The total number of extra hind limbs among the nine specimens examined was 40; one frog possessed 10 hind limbs, whereas five other frogs had seven or more. In many cases, the limbs were poorly developed

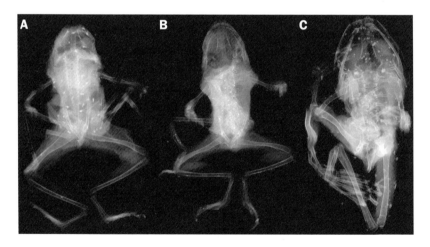

FIGURE 3.14   *Rana catesbeiana*. Specimens collected from a site near Ripley, Ohio, and currently held in the collection of the Hefner Museum at the University of Miami in Oxford, Ohio. Animals were sent to the author for radiography by Curator Michael Wright after Pieter Johnson identified their existence. An additional animal from this site is shown in Figure 2.47. (A) 27 mm SUL; (B) 27 mm SUL; (C) 34 mm SUL.

and completely independent of the appendicular skeleton, originating ventrally from abdominal tissue between the pelvic and pectoral girdles." Radiographs of seven of these animals show no evidence of limbs arising from abdominal tissues, although some limb elements arise from the rostral ilium, which can superficially resemble this condition. Johnson and his colleagues continue in the next paragraph: "Trematode metacercariae were found in preserved amphibians from each site except Ripley Pond. We directly identified *Ribeiroia* metacercariae from five of the six remaining sites . . ."

AN UNUSUAL CALIFORNIA SITE

Pieter Johnson found a site in Santa Clara County, California,[4] that hosted bullfrogs containing *Ribeiroia* metacercariae (Fig. 3.15). These animals had multiple malformations, including bent long bones and unresorbed tails.

AN UNUSUAL INDIANA SITE

A Switzerland County, Indiana, site discovered by Chuck Facemire[4] hosted a population of bullfrog tadpoles and newly metamorphosed juveniles that exhibited three malformation types: an unusual form of polymely, where both hindlimbs were duplicated; hygromas, or subcutaneous, fluid-filled expansions that extend from joint-to-joint (i.e., hip to knee, knee to ankle); and unresorbed tails (Fig. 3.16). The year following the collection of these animals, Facemire returned to the site and could find no frogs at all.

NATIONAL WILDLIFE REFUGES

In 1997, Kathy Converse and her colleagues[9] at the National Wildlife Health Laboratory in Madison, Wisconsin, sampled and analyzed malformed frogs collected on National Wildlife Refuges (NWRs), Wetland Management Districts, and National Parks in Minnesota and Vermont. Malformed frogs were collected at 23 of 38 Minnesota sites (61%) and 10 of 17 Vermont sites (56%). In Minnesota, 110 of 6,632 animals (1.7%), and individuals in 8 of 15 species (53%) sampled were malformed. In Vermont, 58 of 2,267 animals (2.6%) and individuals of 7 in 13 species

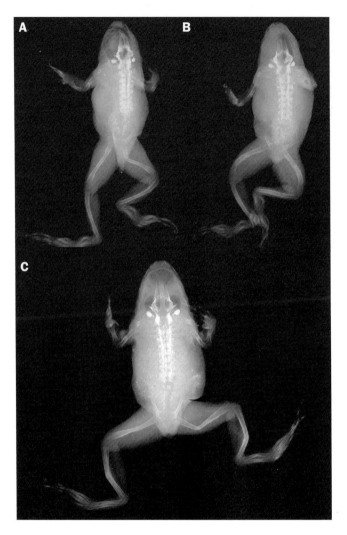

FIGURE 3.15   *Rana catesbeiana*. Collected by Pieter Johnson from a site in California[4] during the 1997–1998 field season. Additional animals from this site are shown in Figures 2.3, 2.29B, 2.32, 2.35, and 2.51. (A) 42 mm SUL; (B) 40 mm SUL; (C) 42 mm SUL.

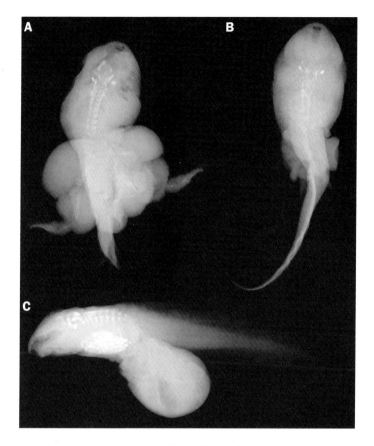

FIGURE 3.16   *Rana catesbeiana*. Collected by Charles Facemire from
Switzerland County, Indiana.[4] (A) ~26 mm SUL. Polymelia + hygroma.
(B) 23 mm SUL. Hygroma. This animal is at an earlier developmental
stage than the animal shown in (A). Two hindlimbs are present and
exhibit hygromas. In addition, two pairs of hygromas, one large and one
small, are present along the ventrolateral edge of the body-tail junction.
(C) ~22 mm SUL. Hygroma. This animal is at an earlier developmental
stage compared with animals shown in (A) and (B); it has not yet formed
hindlimbs. Note the large hygroma, about the size of the tadpole's
body, emerging from the ventral surface of the body-tail junction. An
additional animal from this site is shown in Figure 2.48.

(54%) sampled were malformed. Percentages of malformed individuals/ species ranged as high as 5.2% in Minnesota and 15.6% in Vermont. Most malformations were ". . . partially or completely missing hind limbs or dig-its (50%) or malformed hind limbs and digits (14%). A few individuals had an extra limb or toe, missing or malformed front limb, missing eye or malformation of the mandible" (p. 160). These results raised enough concern that U.S.F.W.S. biologists organized a broad-scale program to sample NWRs across the country. For the past several years I have been receiving malformed frogs from every U.S.F.W.S. region of the country except Region 2 (Arizona, New Mexico, Texas, Oklahoma). Other than the Kenai, Innoko and Tetlin Refuges in Alaska, which have produced high percentages of malformed frogs,[10] few NWRs can be considered hotspots.[11]

FOUR

CAUSES

And now remains
That we find out the cause of this effect—
Or rather say, the cause of this defect,
For this effect defective comes by cause.
Thus it remains, and the remainder thus.

WILLIAM SHAKESPEARE[1]

The year 2006 marked the 300th anniversary of known published observations of malformed frogs. The recognition that malformed frogs preceded the beginning of the Industrial Revolution by about 150 years and preceded modern agricultural techniques (including the application of pesticides) by about 250 years has led many people to deduce that there are natural causes of frog malformations. Indeed there are.

NATURAL CAUSES

In his treatise on the history of malformed amphibians, Martin Ouellet[2] lists several natural causes of malformations.

Wounding   Wounds from failed predation attempts ("bites and mutilations") and leech (*Erpobdella octoculata*) parasitism can lead to either missing limbs or parts of limbs, or polymely through "hyper-regeneration following wounding." See also the comments of Johnson and colleagues,[3] and the discussion of predation in Chapter 5.

Fish excrement   In the European frog *Rana esculenta*, "anomalie P" consisting of bilateral brachymely, polymely, and other types of limb malformations can be induced by rearing tadpoles in the presence of excrement from eels (*Anguilla* sp.) and/or minnows (*Tinca* sp.).[4] Other amphibian species appear to be resistant to these effects. It is unknown whether chemicals contained in fish excrement or microbes transmitted through fish excrement are the cause.

Extreme tadpole densities   Some *Rana esculenta* raised in high densities of 3.5 to 11.1 tadpoles/liter exhibit forelimb ectromelia and hindlimb hyperextension thought to be caused by teratogenic properties of chemical(s) released by the crowded tadpoles. Extreme tadpole densities may also produce wounding.

Lathyrogens   Extracts of sweet pea (*Lathyrus odoratus*) seeds have teratogenic effects on both salamander larvae and frog tadpoles. When young embryos are exposed to lathyrogens, notochord and tail malformations are produced; when older tadpoles are exposed, joint and limb dislocations occur.

Nutritional deficiencies   Malformations produced by nutritional deficiencies are often skeletal malformations, including decreased bone density, scoliosis, joint dislocations, mandibular malformations, and folding fractures of long (limb) bones. Paralysis may also be present. Nutritional deficiencies are most often seen in captive animals. Because most frog tadpoles are herbivorous, and aquatic plants are abundant in healthy, established wetlands, it is difficult to envision food limitation in these ecosystems.

Ultraviolet-B radiation   Andy Blaustein and his colleagues[5] suggest that UV-B can cause severe structural malformations in early-stage embryos of native amphibian species. In the laboratory, Gary Ankley and his colleagues at the EPA found UV-B effects that included bilaterally truncated hindlimbs.[6] The effects of UV-B acting either alone or in association with other factors are interesting and potentially important. I discuss the role of UV-B in producing amphibian malformations in more detail below and in Chapter 5.

Disease   Ouellet cites evidence for production of bilateral posterior ectromelia and hemimely by disease. The potential role for disease in producing malformations must be pursued more ambitiously.[7]

Temperature extremes   Amphibian malformations have been induced by artificially high temperatures (30°C) in the laboratory and naturally low temperatures (cold spring water) in nature. It is unlikely that severe extremes in temperature are experienced by tadpoles in most natural wetlands. However, surface pockets of hot water are present at the NEY pond in Minnesota (Fig. 3.2), and several of the Minnesota wetlands we sampled were spring fed and noticeably cold (CBA, Fig. 3.7).

Hereditary factors   Ouellet notes that some types of malformations are due to genetic mechanisms. However, both Ouellet and I question how similar genetic mutations can arise simultaneously across large geographic regions, or arise simultaneously in several species from the same wetland when animals in the same species from adjacent wetlands are normal.

Parasitic cysts   Pieter Johnson, Kevin Lunde, and their colleagues[8] report a range of malformations induced experimentally by the trematode parasite *Ribeiroia ondatrae* and malformations associated with encysted *Ribeiroia* metacercariae in field-collected animals. Understanding the role of parasites in inducing malformations is critical to understanding the recent frog malformation phenomenon, and with this in mind, I discuss this role of parasites in detail below and in Chapter 5.

MANMADE CAUSES

In addition to these natural causes of amphibian malformations, Martin Ouellet[2] lists several manmade causes of amphibian malformations.

Acidification   Ouellet notes that skeletal malformations such as brachymely and hyperextension may occur when tadpoles are raised in acidified waters. In addition to these potential direct effects, there can be indirect effects: pH levels influence the toxicity of some chemical contaminants (see Chapter 5).

Radioactive pollution   Frog limb malformations have been associated with radioactive waste sites in The Netherlands and in the Federal Republic of Germany. Malformations have also been induced in the laboratory in tadpoles reared in rainwater contaminated with radioactive dust.

Ozone depletion   The destruction of the atmospheric ozone layer leads to increased ultraviolet-B penetrance and therefore to increased UV-B intensity at the Earth's surface.

Heavy metals   Several types of malformations, including eye, pigment, and jaw defects, are associated with metal contamination from coal ash and coal combustion wastes.

Retinoids   A number of authors led by Dave Gardiner (see below) propose that amphibian malformations are a consequence of developmental pathways regulated by retinoids (vitamin A and its derivatives). No single cause of amphibian malformations except retinoids can explain the range of malformations observed in the field. I discuss the role of retinoids, or disruption of the retinoic acid pathways, in producing amphibian malformations in more detail below and in Chapter 5.

Agricultural chemicals (pesticides and fertilizers)   As Ouellet points out: "Agricultural herbicides, insecticides, fungicides, and fertilizers are often toxic to non-target organisms, and can cause deformities and mortality in amphibians." Amphibian malformation types known to be caused by exposure to agricultural chemicals include: limb (shortened or missing limbs or limb segments; multiple limbs or limb segments), tail, and axial (scoliosis). Malformations known to be produced by agricultural chemicals are detailed in Chapter 5. Agricultural chemicals are also known to be lethal to amphibians, acting alone or in combination with other factors.[9]

Other chemicals (xenobiotics)   Tadpoles of various species show many types of malformations when exposed to municipal and industrial waste. In addition, some animals develop tumors or tumor-like cysts. Xenobiotics are considered in detail in Chapter 5.

SYNERGISMS AND NEUTRALIZATIONS

The causes listed above, whether natural or manmade, may or may not occur independently of each other. When causes co-occur, interactions between them may be neutral (no effect), positive (effects of causes become additive or synergistic), or negative (factors wholly or partially cancel one cause or combinations of causes). For example, Marie-Soleil Christin[10] and her colleagues have shown that agricultural pesticides suppress the immune system and increase the risk of parasitic infection by nematodes in northern leopard frogs. Working with wood frogs, Joe Kiesecker took this idea one step further[11] by suggesting that *Ribeiroia ondatrae* also take advantage of pesticide-induced immunosuppression to infect and produce malformations.

Dan Sutherland always maintained that parasites were better environmental indicators than amphibians—that he could tell how polluted a wetland was by using an inverse measure of parasite diversity. He wrote (p. 109):[12] "In my 25 years of experience, when an aquatic wild animal population harbors few parasites, it is often an indicator of habitat degradation or alteration, where intermediate host populations (usually free-living aquatic invertebrates) have been reduced and life cycles of parasites are not completed." If Dan was right, there may be situations where pesticide levels lower parasite numbers and perhaps reduce malformation rates. At any rate, I'm betting along with Kevin Lefferty and his colleagues[13] that pesticide/parasite interactions are complex and depend on species involved (both parasite and amphibian), concentrations of animals (both host and parasite), and types of pollutants, plus a number of other factors that would both surprise and not surprise us.

An overlooked paper in the malformed frog literature is the study by Thiemann and Wassersug[14] examining predator (banded killifish, *Fundulus diaphanous*) effects on parasite (*Echinostoma* sp.) infection rates in *Rana sylvatica* and *R. clamitans*. Thiemann and Wassersug note that tadpole activity reduces trematode infection rates and observe that the presence of either predators or parasites reduces tadpole activity, thereby increasing trematode infection rates. They hypothesize that other stressors, such as the pesticide carbaryl[15] or heavy metals,[16] reduce tadpole activity and elicit responses similar to predator avoidance, creating additive or synergistic effects between stressors and parasitic infection rates.

DETERMINING CAUSE FROM MALFORMATION TYPE: THE SEARCH FOR MORPHOLOGICAL SIGNATURES

One of the difficulties in identifying causes of frog malformations is that there appears to be little correlation between known causes of malformation and the abnormality (i.e., the anatomical position of the abnormality, the type of malformation, and the severity of the malformation do not correlate in instances where we suspect we know the cause). The idea that morphology gives clues to malformation causes is powerful, however, and persists. If true, this notion of a correspondence between causes of malformations and malformation types can be a potent diagnostic tool, because hindlimb development occurs days, weeks, or months (depending on the species and local environmental conditions) before metamorphosis, when most malformed frogs are observed. This temporal disparity may be sufficient to allow whatever caused the malformations to leave or die, or be washed out, degraded, or diluted to the point where they are (or it is), undetectable. Less desirable but still useful is the idea that certain malformation types disqualify potential causes.

Early in the investigation of U.S. amphibian malformations, "bony triangles" (taumelia) were seen as indicative of retinoic acid involvement.[17] However, the demonstration that *Ribeiroia ondatrae* metacercariae can also cause this malformation type[18] forced us to abandon the idea that bony triangles were unique morphological signatures for water-born exposure to retinoid or retinoid-like compounds.

Ankley and his colleagues[6] at the EPA showed that exposure of tadpoles to UV-B radiation produces bilaterally symmetrical ectromelias, making this abnormality an indication of high exposure to UV-B. However, bilaterally symmetrical ectromelias as shown by Ankley are rarely found in nature. In this case, there is a correlation between morphology and cause, but the morphology is rare in nature.

Perhaps there will be a breakthrough. If malformation types can be correlated to causes—if morphological signatures can be established—we'll go a long way towards understanding the biological basis of specific aspects of vertebrate development. Fortunately, the malformed frog investigation does not need to wait for these details. As Chapter 7 will show, the solution to the malformed frog problem is largely independent of cause.

FIVE

RESOLUTIONS

There is a crack, a crack in everything.
That's how the light gets in.

LEONARD COHEN[1]

There has been nothing tidy about the malformed frog investigation. Examining the field, some workers have chosen terms like the "complexity of deformed amphibians"[2] or an "eco-devo riddle"[3] as descriptors. At face value this is true, but I gently disagree that we need more data. The data necessary to achieve a solution to this problem are right in front of us, but at the scale of individual research papers they are often hard to see.

Since the publication of Bill Souder's "A Plague of Frogs"[4] (the paperback edition, published by the University of Minnesota Press in 2002, contains an epilogue that builds a bridge to the most recent papers), it is no secret that at least some of the "science" behind the malformed frog investigation has been driven by bitter personality clashes. Rather than considering science as a process of discovery, science becomes a form of competition. The result is that many appropriate papers are never cited (because they are in opposition), or they are cited but quickly dismissed. You can sense Meteyer's frustration with this situation when she and her colleagues[5] write (p. 170):

From the broad range of malformations in *R. pipiens*, it is clear that there is no simple and effective way to . . . explain the complexity of environmentally induced malformations in wild anurans. Although it may be tempting to focus on one type of malformation or one potential cause, it is our opinion that the data do not lend themselves to simplification. In fact, these data are a call to widen our knowledge about environmental causes and conditions that bring about these malformations.

Ed Ricketts recognized this competitive process at a time when he was a plaintiff in a lawsuit. His laboratory on Cannery Row had burned down due to an electrical malfunction in the adjacent cannery, and after "inspecting the legal system with the same objective care he would have lavished on a new species of marine animal" said (pp. xvi–xvii):[6]

> You see how easy it is to be completely wrong about a simple matter. It was always my conviction—or better, my impression—that the legal system was designed to arrive at the truth in matters of human and property relationships. You see, I had forgotten or never considered one thing. Each side wants to win, and that factor warps any original intent to the extent that the objective truth of the matter disappears in emphasis.

Being aware of this problem, we can see through it. And by seeing through it, we can pick apart the data, keeping the good and tossing the bad. So, like Maclean's smokejumpers who have just landed and collected their gear and know they're in for a long, hot night, we will start deliberately, knowing that small things done correctly now, will later save us time, and perhaps something much more important.

THE PROBLEM OF SCALE

In an effort to establish relevance, authors often generalize and expand their conclusions. In doing so, discipline, discretion, and self-restraint are important attributes; features usually but not always demanded by reviewers and editors. When this does not happen, localized phenomena can be perceived as having broader, even universal, relevance. This mismatch in scale between the scope of the data set and the scope of the problem creates the potential for false conclusions. And in the malformed frog literature this potential has been achieved.

TYPE THINKING VERSUS VARIATION THINKING

Prior to Darwin's publication of the first edition of the Origin of Species (1859),[7] scientists and laypersons alike largely felt that species originated from God via special creation (of course, lots of people who trust science with their lives [and therefore use hospitals], but not with their thoughts, still feel this way). In this pre-Darwinian view, there existed an "ideal" type—the created form—for each species, a concept similar to, and derived from, Plato's ideal of beauty and form represented as shadows cast upon a cave wall.[8] In essence, the pre-Darwinian idea was that there exists a model—a shadow on the wall—to represent each species, ideally (and specially) created. Scientists such as Louis Agassiz, who developed their trade prior to the publication of the Origin of Species, treated specimens that were true to type differently than they treated specimens that exhibited substantial variations, which they either ignored or dismissed, or, when deviations were extreme, wrote them up as freaks.[9] Much of the historical malformed frog literature was generated in this way.

Darwin's notion that differences within species provide the raw material to produce differences between species shifted the attention away from viewing organisms as ideals or models, and toward viewing them as variable. Models and variations each have much to offer as a way of viewing the world, and today you will see both approaches being used in university biology departments.

Malformed frogs are often reported by type (see Chapter 2). Types are convenient, especially as bookkeeping devices, but human convenience does not always translate into biological reality. For example, in Figures 2.11–2.20, I present ten examples of unilateral hindlimb ectromely. Take a close look and you will find that none of the radiographic images of the affected limbs of these animals truly resembles any other. Given this range of choices, which example of unilateral hindlimb ectromely should we use as the ideal representation of this condition? A simple question, but completely unanswerable as asked because this is not a "model" issue, it is a "variation" issue. Carry this idea one step farther and ask, as Meteyer and her colleagues[5] did, if missing limbs are all produced by the same cause, why do the ends of some missing bones expand (Figs. 2.12.–2.15, 2.17, 2.19 and 2.20) while some do not (Fig. 2.16)? And why do some expansions present as increases in cancellous bone (Figs. 2.12, 2.15, and 2.17), while others present as increases in compact bone (Figs. 2.13, 2.14, 2.19, and 2.20)?

There are many ways for missing limbs to present themselves, and rather than trying to dismiss this variation under some umbrella explanation couched in type thinking, it would be better to acknowledge this variability. Ed Ricketts complimented W.K. Fisher on having: "too much integrity to make things fit for convenience what won't fit in fact." This notion of "fit for convenience what won't fit in fact" is probably as good a way as any to state what happens when type thinking inappropriately enters the world of variation. There are many forms of type thinking represented in the variable world of malformed frogs.

THE ADAPTATIONIST PARADIGM

In their seminal paper "The spandrels of San Marco and the Panglossian paradigm: A critique of the adaptationist program,"[10] Steven Jay Gould and Richard Lewontin take biologists to task for their penchant for breaking organisms into "unitary traits and proposing an adaptive story for each considered separately." Gould and Lewontin then expand their criticism to include biologists' unwillingness to consider alternatives to their adaptive stories, for their reliance upon plausibility alone as a criterion for accepting speculative tales, and for their failure to consider competing themes. These competing themes, Gould and Lewontin point out, "are usually dismissed as unimportant or else, and more frustratingly, simply acknowledged and then not taken to heart and invoked." And in a passage that is particularly relevant to the malformed frog phenomenon (p. 585):

> In natural history, all possible things happen sometimes; you generally do not support your favoured phenomenon by declaring rivals impossible in theory. Rather, you acknowledge the rival but circumscribe its domain of action so narrowly that it cannot have any importance in the affairs of nature. Then, you often congratulate yourself for being such an undogmatic and ecumenical chap.

One of the best (but not the only) examples of adaptationist thinking in the malformed frog literature can be found in pages 259–262 in a paper by Stopper, Sessions, and colleagues.[11] My favorite passage here is (p. 262):

> The front limbs of anuran tadpoles develop within the branchial chamber (atrium), which protects them from cercarial attack. The forelimbs of most anurans are used in amplexus, and so are directly related to reproductive success. It is tempting to speculate that atrial development constitutes an adaptation to protect the developing forelimbs from attacks, such as those from trematodes.

Aside from the fact that the best forelimbs ever developed will not be used in amplexus if the hindlimbs cannot get the animal to the breeding site, it is important to understand that among parasites only *Ribeiroia ondatrae* metacercariae are known to cause amphibian malformations, that this parasite has only recently become a problem (an emerging infectious disease in the opinion of some), and that this parasite occurs primarily in North America. The basic tadpole body plan is present globally, wherever there are frogs, and is older than North America, dating back to the supercontinent Pangaea.[12] Therefore, Stopper, Sessions, and their colleagues are asking us to believe the tadpole body plan evolved "adaptations" that would be useful 200 million years in the future and suited only for a subset of species in a fairly limited region of the globe. For a contrasting view, see the recent review of the adaptive, developmental, and genetic underpinnings of frog and tadpole morphology by Handrigan and Wassersug.[13]

Gould and Lewontin also write[10] (pp. 587–588):

> We would not object so strenuously to the adaptationist programme if its invocation, in any particular case, could lead in principle to its rejection for want of evidence . . . if it could be dismissed after failing some explicit test, then alternatives would get their chance. Unfortunately, a common procedure . . . does not allow such definable rejection for two reasons. First, the rejection of one adaptive story usually leads to its replacement by another, rather than to a suspicion that a different kind of explanation might be required. Since the range of adaptive stories is as wide as our minds are fertile, new stories can always be postulated . . . Secondly, the criteria for acceptance of a story are so loose that many pass without proper confirmation. Often, evolutionists use consistency with natural selection as the sole criterion and consider their work done when they concoct a plausible story. But plausible stories can always be told . . . what good is a theory that cannot fail in careful study . . .?

Gould and Lewontin then cite Darwin, who wrote in the last edition of the Origin of Species (1872; p. 395): "Great is the power of steady misinterpretation."

Finally, Gould and Lewontin[10] bolster their central argument by further invoking the spirit of Darwin (p. 590): ". . . we should cherish his consistent attitude of pluralism in attempting to explain Nature's complexity."

It is easy for lay people and scientists alike to confuse anecdote with data. While the word "anecdote" is defined as a short personal account, and synonyms include "story," "tale," and "yarn," the word "data" is defined in part as information obtained from experiments or surveys. Most malformed frog datasets have a lot of anecdote masquerading as data. For example, historical data are largely anecdotal. Field biologists noted curious (usually multilegged) frogs, collected and preserved them, and wrote a short note in a natural history journal. They made little mention of other malformations, and it is not known whether other types of malformed frogs existed historically or whether collection was biased by interest, with missing legged animals stirring little curiosity, perhaps because of the assumption that missing limbs are caused by accidents or failed predation (as they often are, especially in adults).

Modern datasets are equally flawed. Malformed frogs are typically sampled by a biologist, or a crew of biologists, showing up at a wetland at around the time of year when frogs are metamorphosing, and walking or wading the shallow margin, collecting frogs individually with dip nets. Some astonishing numbers and percentages of malformed frogs can be collected in this way at the hottest of the hotspots, and these are the numbers we normally hear (e.g., "The CWB site has a malformation rate of 60%."). But does this technique represent a true (i.e., unbiased) survey? Some considerations come to mind:

1. Would the numbers be the same if the wetland had been sampled at a different time, say a day or two earlier or a week or two later? In his table 1, Merrell[14] writes that on one day, 17 out of 56 northern leopard frog juveniles collected were malformed (30%), while two days later only 7 out of 83 were malformed (8%). Meteyer and her colleagues point out (p. 168):[5] "At one of our study sites in Minnesota (CWB), frogs with polymelia were found in August of one year, but not until October of the second year." Helgen and her colleagues[15] note: ". . . overall malformation frequencies observed can change dramatically in some sites, up or down, from July to late September or early October." Johnson and his colleagues note (p. 348):[16] ". . . dramatic

differences in the patterns of morphological abnormalities in [metamorphosing] amphibians between seasons, ponds, and among seasons and life history stages." Limited sampling times create bias.

2. Do the animals being sampled truly represent the animals that were in the process of metamorphosing, or were numbers reduced by predators? If the latter is true, even though biologists may be sampling in a representative way, the animals remaining might not be representative, since—as Sessions and Ruth point out (p. 46)[17]—the least mobile (i.e., most severely malformed) frogs would be most vulnerable to predation, and the dataset would be biased in favor of healthy frogs.

3. Do the animals being sampled truly represent all the animals that were in the process of metamorphosing, or did the healthy animals move beyond the wetland fringe, while the malformed frogs, being less mobile, stayed behind? If so, the dataset would be biased towards sampling malformed frogs.

4. Do malformed and normal frogs have an equal chance of being captured by biologists using dip nets? If many malformed frogs are indeed less mobile than normal frogs (Helgen and her colleagues note [p. 290]:[18] "malformed frogs . . . were much easier to capture than the normal young frogs"), they undoubtedly have a greater chance of being captured using one-on-one sampling techniques. This creates a bias towards collecting malformed frogs.

5. Does every animal metamorphose? Schotthoefer and her colleagues note (p. 1148):[19] "R. ondatrae . . . infections acquired at the pre-limb bud stage negatively affected [Rana pipiens] tadpole survivorship. In addition . . . the proportions of tadpoles that died at this stage increased with the number of cercariae to which tadpoles were exposed." If some animals are so severely malformed that they cannot metamorphose, or if their development is so delayed that they metamorphose far later in the year than biologists would consider sampling, this creates a bias towards collecting normal frogs.

6. Animals with malformations that affect vital organs likely do not live past embryonic stages, or perhaps past larval stages that require

feeding. These animals never get sampled and this creates a bias favoring normal frogs.

7. Over and over, colleagues and I observe that malformed animals are not as tolerant of stress as normal animals (for example, they often do not survive the usually harmless experience of being put into moist containers in a cooler and transported back to a lab). They are therefore likely to be more vulnerable to naturally stressful conditions, such as high afternoon water temperatures or low nighttime dissolved oxygen conditions, than are normal frogs. Stressed frogs that succumb are never sampled (except as carcasses) and this creates a bias towards normal frogs.

Knowing these potential biases exist allows a better understanding of malformed frog data. For example when 25 multilegged animals are collected from wetland X, we know for sure that wetland X produced 25 multilegged animals, and in truth, that is about all that can be said. Wetland X may have produced 250 multilegged animals but the others never reached metamorphic stages, or reached metamorphosis but got eaten, or metamorphosed but on a day that was not sampled. Wetland X may have also produced missing-legged frogs, but for the same reasons listed above, they never got sampled. Similarly, when an author writes that 60% of the animals at wetland X were malformed, what the data really mean is that 60 of the animals captured at that particular time on that particular day by those particular people using that particular technique were malformed. There has never been a malformed frog field survey that continuously sampled embryos, tadpoles, and newly metamorphosed individuals. This would be an enormously time consuming task, yet it is the only true way to know what percentage of frogs at any particular site are malformed; every other method is a form of shorthand and therefore prone to error.

MALFORMED FROG TYPES

There has been an intellectual bias in the malformed frog investigation, and to illustrate, I use the example of human malformations. Suppose someone stated they were interested in human congenital defects and then focused only on causes of club-feet and polydactyly of the foot. Scientists and clinicians interested in dwarfism, cleft palates, atrial septal defects ("holes in the heart"), hydrocephalus, trisomy 21 (Down syndrome), anencephaly, spina bifida, or even polydactyly of the hand would say, "Yes, that's fine, but your causes do not explain the defects we are interested in and have little relevance to us. Instead, what we are interested in is a broad theory that encompasses all congenital malformations, including those that most interest us."

It is strange that this sort of narrow approach to generalized problems happens all of the time in the malformed frog literature, without the protest. Most malformed frog papers focus on hindlimb defects, because hindlimb defects predominate. But hindlimb problems are not the only malformation types found at most hotspots, and the explanations for hindlimb problems must also take into account the other malformation types. "Solve" the hindlimb issue and you have not solved the malformed frog problem—a large number of other malformation types will continue to cause trouble. Johnson and his colleagues'[20] (p. 163) dismissal of other types of malformations as the result of ". . .'normal' levels of trauma, predation, and developmental error" does not hold for eastern amphibian species, nor for Johnson and colleagues' western toads (see Tables 5.2 and 5.3, below). Again, any general explanation of amphibian malformations should take into account all malformation types; yet by focusing exclusively on hindlimbs (by breaking organisms into "unitary traits and proposing an adaptive story for each considered separately" as pointed out by Gould and Lewontin[10]) they almost never do. Below I list the titles of several of the most recognized frog malformation papers, with the structures considered in bold text:

The effect of trematode infection on amphibian **limb** development and survivorship;

Ribeiroia ondatrae (Trematoda: Digena) infection induces severe **limb** malformations in western toads;

Parasite infection and **limb** malformations: a growing problem in amphibian conservation;

Morphological cues from **multilegged** frogs: are retinoids to blame?

Environmentally induced **limb** malformations in mink frogs;

Hindlimb deformities (ectromelia, ectrodactyly) in free-living anurans from agricultural habitats;

Hindlimb malformations in free-living northern leopard frogs from Maine, Minnesota, and Vermont suggest multiple etiologies;

Explanation for naturally occurring **supernumerary limbs** in amphibians;

How trematodes cause **limb** deformities in amphibians;

Limb deformities as an emerging parasitic disease in amphibians.

Only Dave Gardiner's group has offered a single explanation for nearly all amphibian malformation types found in nature (but see "Retinoic Acid, Other Chemicals, and Chromosomal Damage," p. 139). Other researchers offer combinations of causes, each of which produces a different set of malformations and *together* explain much of the range of malformations found at particular sites. But Gardiner notes, parsimony and strong inference suggest that where you have the option between one and several causes for any particular phenomenon, the simplest option (one cause) will likely be correct. He and his colleagues write (p. 2271):[21]

> The concept of parsimony dates back to the Middle Ages where its roots may be found in the writings of the English philosopher and theologian William of Ockham (1285–1349). His famous principle, often referred to as "Occam's razor," holds that in attributing an explanation to a set of observations, one should not make more assumptions than the minimum needed. The fundamental truth is that when considering a set of models to explain an observation, the **simplest** explanation that explains **all** [bold text theirs] of the observations is most often correct. Occam's razor metaphorically shaves off extraneous assumptions and variables that are not needed to explain the phenomenon in question.

In modeling nature, parsimony denotes maximal explanatory power with minimal model complexity.[22]

Parsimony becomes an important factor when considering the pattern of hotspots on the landscape. However, nature does not always follow the pathways of human logic, and we should take care not to be too accepting or too dismissive based on the way we think the world should operate. As Steve Asma points out (p. 218):[23] "Logical possibility is *not* equal to empirical observation. The former cannot be used to negate the latter; it can only be used to keep our minds open."

LANDSCAPE

Norman Maclean observes:[24] "If you don't know the ground you are probably wrong about nearly everything else." If we believe this, it follows that a clear understanding of the pattern of malformed frog hotspots is essential to understanding the malformed frog phenomenon. Across the western United States, it appears that hotspots are more dispersed than they are in the east. In the east, hotspots are more prevalent in the Upper Midwest, St. Lawrence River Valley, and New England. Maps of malformation sites can be found at the North American Reporting Center for Amphibian Malformations (NARCAM).[25] Two cautions, however, about reading too much into the NARCAM maps: (1) reports of hotspots and reports of single malformations are mapped as single dots, so these maps can diminish the representation of hotspots: (2) historical reports were not peer reviewed, although current reports do receive appropriate peer review.

Having said this, the distribution of hotspots *within* regions is much more important that the distribution of hotspots *across* regions. Within regions, for example the Upper Midwest, hotspots tend to be isolated wetlands; the next hotspot may be four counties away. Hotspots are usually surrounded by wetlands that are normal, and in the Upper Midwest these nearby wetlands are generally well within the home range of individuals of all frog and toad species (Figure 5.1).[26]

Imagine a theoretical landscape, knowing that a variety of malformation types are found at each hotspot, that several species can be affected at each hotspot, and that hotspots are surrounded by wetlands that produce normal frogs. Figure 5.2 shows how Sessions and Ruth's[17] hypothesis—that multilegged frogs are caused by trematode parasites and that missing legged frogs are caused by failed predation—can produce the pattern represented by Figure 5.1. Figure 5.2A shows the wetlands where the density of *Ribeiroia* is high enough to cause malformations (extra

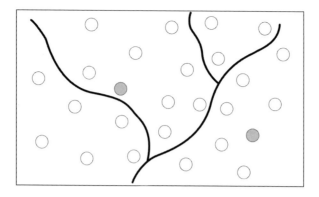

FIGURE 5.1   Hypothetical landscape with wetlands indicated as circles and hotspots indicated by dark-gray circles.

limbs). Figure 5.1B shows the wetlands where the concentrations of predators are high enough to produce missing limbs. Together, then, Figures 5.2A and B produce Figure 5.2C, which reproduces Figure 5.1. What is the likelihood of increased parasites and increased predators co-occurring in those wetlands and only those wetlands? Most would concede that it's small.

What happens if there are regions where parasites are in high concentrations, populated with animals that have extra limbs (Fig. 5.3A), and regions where predators are in high concentrations, with animals having missing limbs (Fig. 5.3B), and they overlap (Fig. 5.3C)? In this scenario (Fig. 5.3C), there are hotspots that host animals with extra limbs and animals with missing limbs. In addition, adjacent wetlands also support malformed animals, but not the range of malformation types that we typically observe at sites in nature (see below). While this landscape is interesting to consider, we know of no hotspot groupings anywhere on Earth that conform to this pattern.

Finally, it might be the case that there are wetlands that, for whatever reason (biological, chemical, physical), have a factor that predisposes them toward creating malformations (Fig. 5.4A). Imagine some regional influence, such as increased UV-B, temperature change, or pH change (Fig. 5.4B), that then triggers this factor to produce hotspots (Fig. 5.4C). For example, the predisposing factor might be chemical (e.g., the presence of a pesticide) and the regional influence might affect breakdown products. There is every chance that this scenario exists in nature.

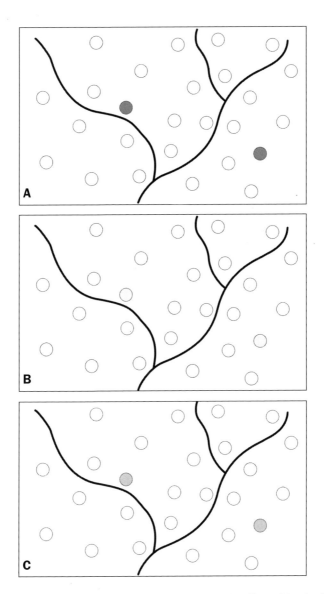

FIGURE 5.2   Hypothetical landscape with wetlands affected by high densities of *Ribeiroia* parasites indicated by black circles (A), wetlands with high densities of predators indicated by light-gray circles (B), and hotspots indicated as dark-gray circles (C). Note that C reproduces Figure 5.1.

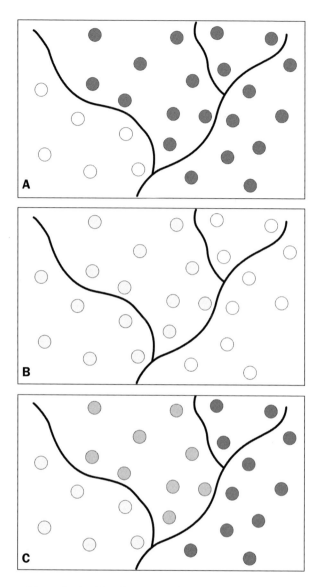

FIGURE 5.3   Hypothetical landscape with wetlands affected by high densities of *Ribeiroia* parasites indicated by black circles (A), wetlands with high densities of predators indicated by light-gray circles (B), and hotspots indicated as light- and dark-gray circles (C).

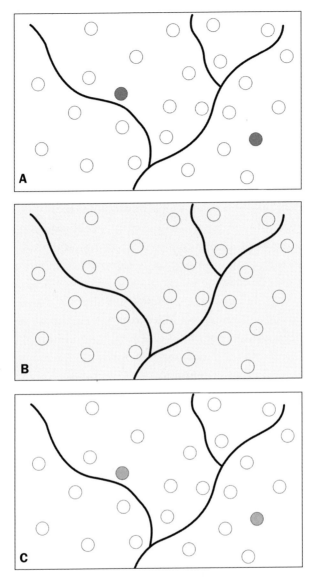

FIGURE 5.4   Hypothetical landscape with wetlands affected by some factor that predisposes them towards creating malformations indicated by black circles (A), and wetlands affected by some regional influence that triggers this factor indicated by light-gray shading (B), that then produces the hotspots as indicated by dark-gray circles (C).

Martin Ouellet[27] notes several important trends in the malformed frog literature. The first is that historical reports of malformed frogs consist predominantly of animals with multiple limbs or limb segments (polymely; again, this pattern may be due to collecting biases of human beings). Second, the number of frogs reported is strongly bimodal, with reports of either one malformed individual, or of >10 malformed individuals, predominating. Third, and building on the second observation, historical accounts of malformed frogs consist mainly of single individuals, while more recent accounts tend toward large numbers of malformed individuals.

To illustrate this third point, Ouellet's data on numbers of malformed frogs per site are plotted against the date of the publication (Fig. 5.5). This graph shows that all sightings of large numbers of malformed frogs have been made since about 1950, suggesting that this problem is historically recent.

A second way to examine Ouellet's third conclusion is to code all reports of malformed frogs in a binary fashion, with fewer than ten individuals as one category and reports of ten or more individuals as a second, and plot them against publication date (Fig. 5.6). Historically, we find scattered "fewer than ten"; more recently, in the 1940s through the 1990s, we find a larger number of "fewer than ten". Beginning about 1950, "ten

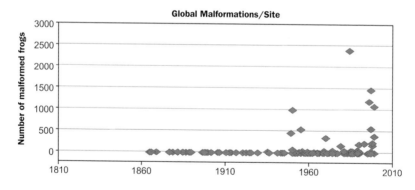

FIGURE 5.5   A plot of Martin Ouellet's global data on numbers of malformed frogs per site against the date of the publication. Although the year of publication may not necessarily correspond to the year when malformations were found, I assume that the lag time from finding to publication is typically not more than about five years, and will not alter the conclusion.

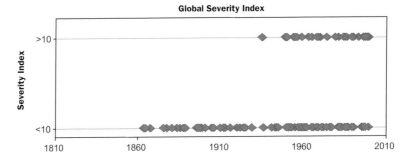

FIGURE 5.6   A plot of Martin Ouellet's data with all reports of malformed frogs with "fewer than ten" individuals grouped together and
reports of "ten or more" individuals grouped together, and plotted
against publication date. Note that historically, we find scattered "fewer
than ten"; more recently, in the 1940s through the 1990s, we find a
larger number of "fewer than ten" reports. Beginning about 1950, "ten
or more" appear and continue, along with the "fewer than ten" to the
present day. These reports of "ten or more" constitute the malformed
frog problem.

or more" appear and continue, along with the "fewer than ten", to the present day. These reports of "ten or more" can be said to constitute the malformed frog problem, although there is a long lag time between the earliest
reports of multiple malformed frogs per site and 1995, the time when society became aware and alarmed, and the problem became a phenomenon.

There is an additional trend that Ouellet likely observed but did not
note— that the frequency of literature reports of malformed frogs has increased (Figs. 5.5, 5.6). It is hard to know what to make of this. This
might, and probably does, reflect a real phenomenon (what most people
think, see for example Blaustein and Johnson's articles[2, 28]). It could also
be an artifact of modern science—the overall number of published scientific papers is increasing and malformed frog papers might simply reflect
this scientific activity. It could also be an artifact of modern society. There
are more people, and people are more mobile with more access to rural
places than at any time in the Earth's history. Frequency of reported malformed frogs might simply reflect the same number of malformed frogs
as always, but a higher rate of encounters with humans who have learned
to be concerned. Finally, it is likely that while historic workers simply

stumbled onto malformed amphibians, modern workers actively seek them out, and this switch from a passive to an active search pattern could be reflected in additional reporting.

The same patterns of malformation occurrence hold for the United States. Repeat Ouellet's analysis—examine the number of malformed frogs against the date of the study—using only data from the United States and you again find that large numbers have been reported only recently, and that five of the seven reports of large numbers have been published since 1999. Group the studies according to whether fewer than ten individuals were reported or ten or more individuals were reported and you see the same pattern for the United States that you see globally—that all reports of large numbers of malformed frogs have been published since 1950.

Ouellet's analysis is complemented by Dave Hoppe's[29] findings on the history of malformed frogs in Minnesota. Hoppe compares his findings to those of previous workers, including the museum specimens that formed the dataset for Dave Merrell's (1969) paper,[14] and asks "Is there anything new?" There is. Hoppe found that recent malformation rates were about ten times higher than in Merrell's dataset. Hoppe also found that the two most common malformations in his sample—hindlimb skin fusions and corkscrew-shaped limbs (anteversions)—were never found in Merrell's frogs, despite being malformation types easily noticed in the field. Several other malformations Hoppe found in the field, including split limbs, missing eyes, malformed jaws, missing forelimb elements, and hip dysplasia, were also never found in museum specimens. Instead, museum specimens exhibited missing limbs and missing, fused, and multiple digits. Hoppe[29] concludes: "Recent findings of anuran abnormalities in Minnesota . . . represent a new phenomenon. Frog abnormalities were more frequent, more varied, more severe, and more widely distributed in 1996–1999 than in 1958–1992."

Amphibian malformations are probably more numerous, and likely more severe, in modern times than they were historically. Does the malformed frog phenomenon, therefore, constitute a "spreading epidemic (p. 65)"?[22] According to Stedman's Medical Dictionary (24th edition), epidemic means: "1. A disease attacking many in a community simultaneously; distinguished from endemic, since the disease is not continuously present but has been introduced from outside [Johnson and his colleagues write (p. 377):[30] ". . . R. ondatrae is not an invasive trematode in North

America"]. 2. A temporary increase in number of cases of an endemic disease." This second definition may in fact be true; every limnologist knows that the key to low nuisance algal levels is drought (less runoff means fewer nutrients, which means fewer, smaller plants). Therefore we might expect waxing and waning of parasite-induced malformations (less plant life means fewer snails, and therefore fewer primary intermediate hosts and fewer *Ribeiroia*), corresponding to hydrologic cycles. However, waxing and waning, even if it is occurring on an increasing slope as generally more and more nutrients get pumped into ecosystems, is not the same thing as a "spreading epidemic." Further, Johnson and his colleagues'[31] comparison of historical with current data from the Ripley, Ohio site show that the malformation process is reversible. This pond, which in the middle 1950s produced many animals with some of the most bizarre malformations ever radiographed (see Figs. 2.47 and 3.14), is today not considered a hotspot.

DO MALFORMATIONS CONSTITUTE DISEASE?

In their 2003 paper in Conservation Biology, Johnson and his colleagues write (p. 1727):[31] "Here, we classify parasite-induced malformations in amphibians as a disease, recognizing the gross morphological alterations and elevated mortality resulting, directly or indirectly, from infection." They cite an example supporting their position (with the word ectromelia misplaced): "Other pathogens, such as mousepox virus, (ectromelia) for example, are also known to induce limb abnormalities within infected hosts."

The question is, do frog malformations constitute a "disease" or not? As Johnson and his colleagues state (p. 1727):[31] "Defined medically, a *disease* [italics theirs] represents any deviation from or impairment of the normal structure or function of any part, organ, or system of the living animal or plant body (Merriam-Webster, 1997; U.S. Food and Drug Administration, 1999)." People have questioned whether definitions in Merriam-Webster and by the U.S. F.D.A. constitute a medical definition (one person noted that since the U.S. F.D.A. does not define alcohol as a drug, U.S. F.D.A. definitions are as much political as they are factual, and a better governmental definition might come from the CDC [Centers for Disease Control]). I'm not sure how much difference this makes. Stedman's defines disease as: "1. Morbus; illness; sickness; an interruption, cessation, or disorder of body functions, systems, or organs. 2. A disease entity characterized usually by at least two of these criteria: a recognized etiologic agent (or agents), an identifiable group of signs and symptoms, or consistent anatomical alterations." Johnson and colleagues could have substituted this definition for the one they used and still made the same argument.

Johnson and his colleagues continue:[31] "Clinical signs of a disease may be morphological, physiological, or behavioral, and the etiological agent may be infectious, genetic, nutritional, toxicological, or traumatic." Under this definition, gunshot wounds, injuries from car accidents, and deaths from hurricanes constitute disease, which may strike us as strange. But that is the great thing about working in an area where creative people care deeply—they say and do things that make you think. While most of us, including a fair number of epidemiologists, do not consider automobile accident and hurricane fatalities to be disease, the definition is broad, and shares with many Bible verses the ability to be freely interpreted. So when considering the definition of disease we are perhaps left with the approach of Justice Thurgood Marshall when trying to define another subject—"I know it when I see it," or the immortal words of John Prine:[32] "It makes no sense that common sense, it makes no sense no more."

I can accept, with a shrug, the notion that amphibian malformations constitute disease. What I don't understand is why Johnson and his colleagues chose such a narrow definition—". . . we classify parasite-induced malformations as a disease . . ."—when it seems obvious that all amphibian malformations could be classified as disease under the broad definition they use. I realize that this definition does not disqualify other causes of malformations for consideration as diseases, or, more broadly, other diseases. But because this definition is not inclusive, it creates a lot of problems for field biologists trying to understand amphibian malformations. For example, the bullfrog in Figure 2.47 would not be classified by Johnson and his colleagues as diseased, because their own data[31] show that no bullfrogs, either historically or currently, from the Ripley, Ohio, population contain *Ribeiroia* metacercariae. Similarly, while the northern leopard frog shown in Figure 5.7A would be classified as diseased (since it came from the NEY site in Minnesota [Fig. 3.2], which is known to harbor high densities of *Ribeiroia*; see below), an almost identically-appearing northern leopard frog shown in Figure 5.7B would not (since

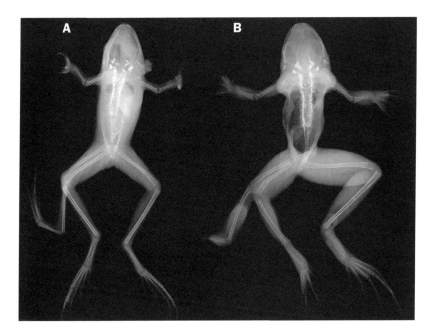

FIGURE 5.7   A comparison of "diseased" (A) and "non-diseased" (B) multilegged malformed frogs according to the definition of Johnson and his colleagues.[31] (A) *Rana pipiens* collected at the NEY site in Minnesota, which carried high densities of *Ribeiroia* (also shown in Fig. 2.36A); (B) *Rana pipiens* from the TRD site, also in Minnesota, which had no *Ribeiroia* (also shown in Fig. 2.36B).

the TRD site [Fig. 3.4], also in Minnesota, was newly created and at the time this animal was collected, had no submergent plant life, no established snail populations, and the animals carried no *Ribeiroia* metacercariae).[33]

Other problems arise. Under Johnson and his colleagues' narrow definition, malformations on different parts of the body may or may not constitute disease. Table 1 in Johnson and his colleagues' 1999 paper[34] (reproduced in part here as Table 5.1, see below) compares malformation types and rates by body region (cephalic/axial, forelimb, hindlimb) in field-collected Pacific treefrogs to malformation types and rates by body region in *Ribeiroia* infected experimental animals. Here, you find that hindlimb malformations, being the only malformations scored as being induced by parasites, count as a disease, while forelimb and cephalic/axial malformations do not.

Further, the same malformation on the same body part may or may not constitute disease depending on the species being considered. Comparing Table 1 in Johnson and colleagues' Canadian Journal of Zoology paper[30] (reproduced as Table 5.2 here, see below) to Table 5.1, you are forced to conclude that, whereas forelimb malformations in western toads constitute disease, forelimb malformations in Pacific treefrogs, many of which are the same malformation type, do not.

DO *RIBEIROIA* METACERCARIAE CLEAR
IN POSTMETAMORPHIC ANIMALS?

One problem hindering a general acceptance of the parasite theory of amphibian malformations is that *Ribeiroia* metacercariae do not appear in animals from all populations, or even in all malformed frogs from *Ribeiroia*-infected populations. It is known that trematode metacercariae will clear from hosts, especially fish hosts, and so one idea, first proposed on an internet listserve, was that *Ribeiroia* metacercariae clear and therefore create malformations without appearing to have been present. Joe Kiesecker follows this up when he writes (p. 9903):[35] ". . . tadpoles possess the ability to prevent the formation of, or to shed, metacercarial cysts . . . it will be important to determine whether metacercariae can be shed after they induce limb deformities. Such a pattern could explain the occurrence of deformities in wild collected frogs that are not associated with metacercariae." Stopper, Sessions, and their colleagues[11] continue this line of reasoning (p. 261): "In past studies, the role of cysts may have been underestimated because of the possibility that frogs may be able to clear cysts (probably due to immunological rejection of the cysts) . . ."

Can *Ribeiroia* cysts create the sort of massive tissue disruptions described by Stopper, Sessions, and their colleagues[11] and, in the case of American toads and other species with short tadpole stages, clear within one or two weeks so as to leave absolutely no trace of their existence? And can there be so much site-specific variation in this clearing process that the results of collections at a large number of sites over a restricted period of time (such as the collecting trip Sutherland and I took; see Minnesota parasite data, p. 163) are due to differential clearing of parasites, not to parasites being present or absent? In fact, these speculations of Kiesecker and Sessions' group are in direct opposition to data collected by Pieter

Johnson's group, who write (p. 232):[36] ". . . numerous [*Ribeiroia*] metacercariae may be recovered from mature frogs that are several years old, suggesting that amphibians may be less effective at eliminating *Ribeiroia* than are fish. These metacercariae are often brown in colour, however, possibly indicative of the colonization of host eosinophils."

In essence, what Kiesecker and Sessions are arguing is that when parasites are present, they are present, and when they are absent, they may have also been present. But when a group such as Brian Eaton's writes (p. 284):[37] "We found three wood frogs with extra hind limbs, a condition that has been linked to the presence of cysts of the trematode *Ribeiroia ondatrae* in wood frogs and other anurans. Two of these specimens were . . . examined for the presence of helminth parasites . . . One of the preserved specimens had no helminth parasites, whereas the other had two metacercarial cysts of the family Echinostomatidae . . . we are confident that no metacercariae were missed"—it is wise to believe that no *Ribeiroia* were present in these animals. The assumption that *Ribeiroia* present means *Ribeiroia* present, and *Ribeiroia* absent means *Ribeiroia* absent is scientifically defensible. Indeed, this perspective becomes extremely useful when we stop considering *Ribeiroia* in isolation, and instead consider the entire trematode parasite fauna infecting these frogs (see Minnesota parasite data, p. 163).

DOES AGRICULTURE CAUSE AMPHIBIAN MALFORMATIONS?

In the late 1990s, Martin Ouellet and his Canadian colleagues published a series of papers[38] on amphibian malformations from agricultural areas, concluding that there is an association between agricultural regions, especially with pesticides from agricultural regions, and malformations. In 2002, Pieter Johnson and his colleagues came to the opposite conclusion and wrote (p. 163):[20] ". . . [our data] suggest that agricultural impacts from pesticides at these sites are either very slight or nonexistent." In 2006, Taylor and colleagues[39] re-affirmed the link suggested by the Canadians when they found an association between malformations and proximity of the affected site to areas with high human land uses, including agriculture and lawns. The pendulum swung back in late 2006, when Piha and colleagues wrote (p. 810):[40] "We found no evidence for increased abnormality frequencies in the habitats most likely exposed to agrochemicals."

It swung back again in early 2007 when Gurushankara and colleagues[41] found malformations associated with "agroecosystems" in the Western Ghats region of India. This debate will likely continue.

The problem here is that the question, "Does agriculture cause amphibian malformations?" is too big for the datasets being collected. Malformed frog hotspots tend to occur as isolated wetlands. Can you find isolated hotspots associated with agriculture? There are dozens in the Midwest. Can you find isolated sites associated with agriculture without large numbers of malformations? Most wetlands in the Upper Midwest fall into this category. In many places across the world, agriculture may form a small portion of the landscape and in these places there is every chance of an association between malformations and agriculture. But in the Midwest, where the malformed frog phenomenon burns the hottest, agriculture *is* the rural landscape, and what is not agriculture is typically either lawn or has been urbanized. Yet, the number of hotspots is relatively small compared to the number of wetlands on the landscape. And so, yes, there is an association between malformations and agriculture, and no there is not. Put another way, no matter which way you answer the question "Does agriculture cause amphibian malformations?" there are plenty of data to argue the contrary position. If you alter the question slightly and ask, "Does agriculture cause amphibian malformations *at this wetland*?" and do not extrapolate, the scale of the question and the scale of the data fall into register.

DECONSTRUCTING THE MAJOR HYPOTHESES
PREDATION

When we hear that missing limbs are caused by failed predation, we know there is an element of truth in this. Tennyson's famous line, "nature red in tooth and claw" includes frogs with limbs that have been bitten off, and it makes sense that missing limbs are caused by failed predation. Sessions and Ruth[17] argued that their parasite hypothesis, which addressed multi-limbed malformations, in combination with amputations due to failed predation attempts, explained amphibian malformation types observed in the field. This view has been widely perpetuated, both among herpetologists and the general public, although to many this scenario is highly dependent on co-incidence (Fig. 5.2). Across the Upper Midwest, through the St. Lawrence River Valley, and throughout New England, most malformed frogs are missing limbs or missing limb segments. There are several problems with the notion that failed predation attempts generally explain these missing limbs:

1. Scarring, or signs of wound repair are rarely observed in metamorphic frogs. Among developing limbs, the older the limb, the longer the limb, and the likelihood of being grabbed by a predator must be in some proportion to its length. It therefore seems that the closer a tadpole gets to metamorphosis, the more likely it is to lose a limb, and the more likely it would be to show obvious signs of wounding or wound repair.

2. Pigment malformations often occur (> 70% in the Minnesota malformed frogs I have examined) on the proximal portions of missing limbs. Why should an amputation convert, say, a barred pigment pattern to a mottled pigment pattern (see Color Plate 4) at some distance from the allegedly traumatized area and in the absence of scarring? Amputation is not known to be accompanied by widespread pigment alterations.

3. Missing arms are often accompanied by missing shoulder girdle elements (Fig. 2.9 and 2.10); missing legs are often accompanied by missing pelvic elements, especially ilia (Figs. 2.1–2.6). This would represent severe trauma, and such trauma cannot be consistent with life.

4. Why would predators selectively choose hindlimbs? Early in development hindlimbs are tiny and tucked under the body and tail. Failed predation attempts on tadpoles by aquatic vertebrates typically result in lost tails. In the prairie pothole region of North America, failed predation by birds on tadpoles in drying wetlands frequently produce a "v" shaped notch, corresponding to beak morphology, in the tadpole's dorsal fin.

5. Developing amphibian limbs cannot be easily amputated by being pulled from the body (as would occur if grabbed by animals with mouthparts adapted for swallowing prey whole [most non-avian amphibian predators]), but instead must be severed by a shearing masticatory apparatus. What kind of predators have this type of shearing masticatory apparatus and attack from below? Aquatic turtles are candidates, but what kind of turtle population would it

take to produce malformation rates approaching 70% in a frog population?

6. If the function of the presence of *Ribeiroia* metacercariae in frogs is to produce (generally multilegged) limb malformations to facilitate predation by animals that serve as definitive hosts for *Ribeiroia*, why do normal frogs become such major targets for predators (in order for so many missing-limbed animals to be produced)? It seems that if multilegged malformed frogs are targets and predator abundance is high, this would be reflected in a reduced number of multilegged malformed frogs, not an increased number of missing-limbed frogs.

7. Missing limbs occur in the absence of predators. Malformed green frogs were observed in 1997 at a newly constructed wetland east of Indianapolis. Most malformations were missing limbs, yet for the first year there were no aquatic predators (fishes, turtles, or invertebrates) at what was essentially a big mudhole. In Dave Hoppe's (personal communication) cage-rearing experiments, done with northern leopard frog and American toad tadpoles, mesh size was 1 mm (3 mm when holding large northern leopard frog tadpoles), animals were raised to metamorphosis, and the common malformation type was ectromelia (same as in the wetland). What sort of predator can negotiate a 1 mm mesh, manage to bite off a frog's leg, eat it, and then escape?

8. Preliminary studies examining the histology of the spinal cord demonstrate that in many animals with missing limbs, motor neurons in the spinal cord are also absent, suggesting that missing limbs and limb segments were never present.[42] This evidence argues for a developmental origin of missing limbs and *against* the argument of failed predation.

9. Predators are typically never identified, or when identified are generally not reported in high enough concentrations, or determined to be present in enough wetlands, to represent a general cause.

Gardiner and his colleagues[21,43] present other circumstantial evidence against failed predation as a leading cause of missing limbs, and today, failed predation is not considered to be a general explanation for missing limbs (see, for example, papers by Meteyer and colleagues,[5] Levey,[44] and Taylor and colleagues[39]). However, in particular wetlands with high densities of predators, perhaps even conspecifics, it is possible for failed predation to produce missing limbs or missing limb segments. These specific situations do not justify broad conclusions about the role of predation in producing malformed frogs.

What is of some concern is the implication of Sessions and Ruth's[17] conclusion that extra limbs are caused by trematode cysts and that missing limbs are caused by failed predation. Several workers, having examined malformed frogs and not found parasites, have concluded that failed predation is the likely cause.[2,28,37] But many other causes of missing limbs are known (see below) and it is certainly unwise to dismiss these outright.

ULTRAVIOLET-B RADIATION

Direct effects   The 1997 paper by Andy Blaustein and his colleagues[45] demonstrates that UV-B can cause severe structural malformations in early-stage embryos of native amphibian species. One consequence of the ways field researchers typically sample malformed frogs (that is, at metamorphosis, after these animals attempt to crawl out onto land) is that they miss sampling early-stage embryos, because these animals do not survive to metamorphosis. Any true measure of the rate and severity of amphibian malformations should include sampling embryos, tadpoles, and newly metamorphosed juveniles. This would be time-intensive; but in the absence of such studies, it is difficult to extrapolate Blaustein and colleagues' results. Doing this work should be a top research priority.

A spatial analysis of known malformation hotspots does little to argue for or against the effects of UV-B. Even within a restricted geographic region, exposure to UV-B varies with tree canopy cover, extent of emergent and submergent vegetation, water clarity, water color, and water depth.

In the laboratory, Gary Ankley and his colleagues[46] examined the role of UV-B and methoprene (a retinoid compound applied to wetlands to control mosquito larvae) and found UV-B effects that included bilaterally truncated hindlimbs. Developmental biologists often comment on how rarely bilaterally symmetrical limb defects are induced, presumably because critical developmental windows are open for only a short time and vary slightly, even between limbs on the same animal (Gardiner and colleagues[21]). The fact that UV-B exposure reliably induces bilaterally symmetrical limb defects is interesting. In all of Ankley and colleagues' experimental animals, the femurs of the foreshortened limbs simply end—there are no bony expansions. Gardiner and his colleagues[21] attribute these features to a simultaneous UV-B induced destruction of apical ridge ectoderm in both limbs.

In the field, bilaterally symmetrical limb defects are rare, and bilateral truncations are rarer. Further, the truncated limbs produced in the experiment of Ankley and his colleagues show simple bone truncation, whereas field-collected animals in samples often have large spongiform bony expansions at the point where the limbs are missing (Fig. 2.21). Blaustein and Johnson[2] point out (p. 61) that increases in UV-B radiation do not explain all types of malformations, even all types of limb malformations, found in nature.

Indirect effects   It is possible for the pattern of hotspots we observe to be due to indirect UV-B effects, with hotspots being sites predisposed to being vulnerable to UV-B exposure due to the presence of some other factor (see Fig. 5.4). Jim La Clair and his colleagues[47] suggest that S-methoprene, a retinoid used to inhibit insect metamorphosis (e.g., from maggot to fly or wriggler to mosquito), poses little risk to *Xenopus* embryos, but that its breakdown products, which are influenced by sunlight, water chemistry and biological activity, cause malformations. Fernandez and l'Haridan[48] demonstrated that UV-B irradiation can also increase the toxicity, and therefore the teratogenic effects, of polycyclic aromatic hydrocarbons. As Tyrone Hayes recently noted:[49] "It's not the pesticides alone or introduced predators or ultraviolet light or global warming that's causing this decline, but the interaction between these on an animal that's pretty sensitive to its environment."

More data on the extent and nature of the role of UV-B in inducing malformations are needed. Unfortunately—unlike birds of prey, or honey bees for that matter—our retinas are not built to detect UV-B radiation. And so important changes (usually increases) in UV-B intensity can occur while we, with the limitations of our human sensory systems, can be completely oblivious.

Retinoic acid Developmental biologists, particularly those interested in limb formation, thought they had a solution to the malformed frog problem: an interesting group of molecules termed retinoids, related to vitamin A. The alcohol form of vitamin A, or retinol, is converted by a barrage of enzymes into active retinoid aldehydes or carboxylic acids and used either for vision or in hormonal signaling processes that activate genes expressed during development. It is this latter function that interests developmental biologists, because, not surprisingly, these genes are crucial for normal development. Retinoid pathways that regulate gene expression are highly conserved in chordate animals, including vertebrates.

Developing systems known to be influenced by retinoic acid–mediated processes include the limbs; the central nervous system (brain and spinal cord); skin; craniofacial features; eyes, including the retina; the circulatory system, including the heart; other internal organs including intestines and kidneys; and even genitalia. Exact quantities of retinoids must be present during critical developmental stages. Developmental defects, ranging from minor to serious to fatal, result from increases or decreases in retinoid signaling (Gardiner and colleagues[21, 43]).

Dave Gardiner and his fellow developmental biologists noted that retinoids could produce the full range of observed malformed frog types, including both multiple and missing limbs, as well as an assortment of defects in other organ systems.[21, 43] They then employed the ideas of parsimony and strong inference to argue for retinoids and against parasites as the cause of amphibian malformations (see Malformed Frog Types, Chapter 2), as follows. Gardiner and his colleagues[21] first dismissed the parasite hypothesis of Johnson and his colleagues because there are no mechanisms known to developmental biology that would allow *Ribeiroia* to have the effects that have been attributed to it. That is, mechanical disturbance alone, as proposed by Sessions and Ruth[17] and Stopper, Sessions, and their colleagues,[11] cannot explain all malformation types attributed to *Ribeiroia* infection, although Gardiner and his colleagues did allow that parasite metacercariae could be producing a retinoid-like substance that affects limb development. Gardiner's group then noted that the Sessions and Ruth hypothesis consisted of two hypotheses (parasites for multiple limbs; failed predation for missing limbs). Because the retinoid hypothesis explains almost all malformation types, while Sessions and Ruth's hypothesis explains only a subset of fairly rare malformation types, parsimony suggests that retinoids are the single cause of amphibian malformations: "Taken together, the laboratory and field observations strongly support our contention that an environmental retinoid(s) is present and responsible for frog deformities in the water samples from the sites tested. We infer that the same compounds will be found at other deformed frog sites . . ." (p. 2268).[21]

Sig Degitz and his colleagues, working with *Xenopus*, came to the exact opposite conclusion. They write (p. 139):[50] ". . . it is unlikely that retinoid mimics would produce the spectrum of limb malformations which recently have been observed in amphibians collected from the field." This finding is surprising because it contradicts the conclusion of their earlier study (p. 264) where Degitz and his colleagues write:[51] "[retinoic acid] exposure at stage 51 in *X. laevis* and stage 28 in *R. sylvatica* resulted in concentration-dependent increases in reductions and deletions of the hind limb." In this same paper they report species-specific differences in response to retinoic acid within genera (*Rana*), and the across families (Ranidae vs. Pipidae). Given this interspecific variation, it is interesting to me that in their subsequent laboratory study, they would use *Xenopus*, which is a non-native species and has a different response to retinoids,[52] to draw conclusions about field data on native *Rana*. Species-specific responses are not restricted to chemical insults; there is a similar response to parasites. Pieter Johnson and his colleagues point out (p. 374):[30] ". . . differences in malformation types and their relative frequencies between Pacific treefrogs and western toads indicate that amphibian species may respond differently to identical numbers of *R. ondatrae* cercariae."

The real problem for the retinoid hypothesis is the lack of strong field data. While Gardiner and his colleagues[21] show the presence of biologically active retinoids from the CWB site in Minnesota, and a site in Mission Viejo, California (their figs. 2 and 3, pp. 2269 and 2270), there have been no published data sets showing a regional relationship between field-levels of retinoids, or chemicals demonstrated to affect retinoic-acid pathways, and malformations; nor has a dose/response curve been generated. In the absence of adequate field samples, the notions of parsimony, or strong inference, as logically appealing as they are, are not sufficient. Gardiner and Hoppe (p. 215) realize this and conclude:[43] ". . . frog malformations arise as a consequence of acute exposures—either a single exposure, or more likely a series of exposures to a teratogenic agent—rather than a continuous, chronic exposure. Sampling protocols that attempt to identify the agent(s) responsible for these malformations need to accommodate this likelihood."

Other Chemicals   In her 1963 classic, Silent Spring,[53] Rachel Carson notes that when you put something in water anywhere, it ends up in water everywhere. This is more true than not, and it tends to be a problem for people interested in water chemistry. Not a problem for the usual reason of trying to find something that may be absent, but for the opposite reason, finding meaning in something that is ubiquitously present. If you ask an environmental chemist about which chemicals occur in a particular water body, he or she will invariably respond with words to the effect: "Tell me what you're interested in and we'll find it."

Complicating matters, chemicals can be ubiquitous at levels that are not detectable by instrumentation. Tyrone Hayes and his colleagues[54] have discovered biological activity in atrazine concentrations approaching the level of our ability to detect atrazine. It is possible that chemicals are having biological effects at concentrations we cannot measure.

To get around this problem of detectability, several labs have exposed frogs to water, or to extracts of water, from malformed frog hotspots. (Usually, and unfortunately, these frogs are *Xenopus*, which is a poor model,[52] because they are not native to North America [or the New World, for that matter] and are phyletically distant from most North American frog species. Furthermore, *Xenopus* do not live in regions with high malformation rates.) These studies always find effects. Joe Tietge's group[55] attributes impacts to the ionic composition of the water. Jim Burkhart's group reports (p. 841):[56] "Initial experiments clearly showed that water from affected sites induced mortality and malformation in *Xenopus* embryos, while water from reference sites had little or no effect. Induction of malformations was dose dependent and highly reproducible . . . Limited evidence from these samples indicates that the causal factor(s) is not an infectious organism nor are ion concentrations or metals responsible for the effects observed. Results do indicate that the water matrix has a significant effect on the severity of toxicity." Doug Fort's group reports (p. 2310):[57] "The results shown here demonstrate that water and sediment samples from several Minnesota and Vermont pond test sites induce adverse development effects, typically maldevelopment . . ." They then used a subset of compounds identified in the first study, including propylthiourea, diphenylamine, bisphenol A, permethrin, desisopropyl atrazine, nickel chloride, aldoxycarb, and Maneb and tested their ability to produce malformations in *Xenopus*.[58] All compounds produced malformations; effects included malformed viscera, notochord defects, craniofacial and eye defects, axial flexures and hindlimb cartilage defects. Further, Fort and his colleagues found (p. 2316): "The potency of several compounds was also enhanced by the site water, thus indicating that the water matrix deserves consideration as a contributing factor for both laboratory and field studies."

Hopkins and his colleagues[59] and Rowe and his colleagues[60] (both in collaboration with Justin Congdon) demonstrated axial and oral malformations in bullfrog tadpoles collected from polluted habitats. Tina Bridges[61] found malformations in southern leopard frogs (*Rana sphenocephala*) induced by exposure to the insecticide carbaryl. Indeed, the field of amphibian and reptile susceptibility to toxins has its own bible, the book by Don Sparling, Greg Linder, and Christine Bishop.[62] (It is curious that the same malformation types used to argue for parasite induction of malformed frogs appear on the cover of this definitive ecotoxicology text.) A careful reading of the chapter by Cowman and Mazanti reveals several compounds known to induce amphibian malformations, including:[63]

the organophosphate insecticide acephate, which produced abnormalities in *Ambystoma gracile* at high concentrations;[64]

the organophosphate insecticides dimethoate and dichlorvos, which produced melanophore (pigment) malformations in *Bufo melanosticus*[65] and *Rana tigrina*;[66]

the organophosphate insecticides malathion, dicrotophos, monocrotophos, parathion, and the metabolites malaoxon and paraoxon, which caused reduced size, abnormal pigmentation, abnormal gut, and abnormal vasculature including the heart in *Xenopus laevis* tadpoles;[67]

the organophosphate insecticide methyl parathion, which caused skeletal, spinal and tail malformations in *Rana perezi* tadpoles exposed to low doses;[68]

combinations of the organophosphate insecticides phenyl saliginen cyclic phosphate, leptophos-oxon, tri-o-tolyl phosphate, and para-oxon, which produced spinal malformations and blisters in three species of native amphibians (*Hyla chrysoscelis*, *Gastrophryne carolinensis*, and *Rana sphenocephala*) representing three different families;[69]

combinations of the organophosphate insecticides Basudin®500EC, diazinon, Dithane®DG, Imidan®50WP, Guthion®50WP, and Nova®40W, which produced mortality, malformations, or delayed growth in *Rana pipiens* and *Rana clamitans*;[70]

the carbamate insecticide carbofuran, which produced mid-trunk region and tail malformations in *Microhyla ornata*[71] and ectromelia and ectrodactyly in juvenile *Rana clamitans*;[72]

the carbamate insecticide oxamyl (granular formulations of which have been banned in the United States[73]), which produced spinal curvature and tail tip malformations in caged *Rana temporaria*;[74]

the carbamate insecticide primicarb (ZZ-Aphox®), which produced notochord and vascular malformations in *Rana perezi* tadpoles and, in combination with the organophosphate Folidol®, produced vertebral column and limb malformations in the same species;[73]

the carbamate insecticide propoxur, which produced pigment malformations and kinked tails in *Rana hexadactyla*;[75]

residues of the pyrethroid insecticide fenvalerate, which produced "bent" backs in *Rana clamitans* embryos;[76]

the herbicide paraquat, which produced shortened and flexed tail malformations in *Rana pipiens*;[77]

the herbicide diuron, which produced limb malformations in *Rana aurora*;[78]

the fungicides chloranil and dichlone, which produced pigment, muscular, sensory system, other cephalic, and "body shape" malformations in *Xenopus laevis*;[79]

the fungicide manganese ethylene bisdithiocarbamate (MANEB), which produced skeletal abnormalities including limb malformations in *regenerating* [italics mine] limbs in the newt *Triturus cristatus*;[80]

the fungicide tributyltin oxide, which induced skeletal malformations, including limb malformations, in *Ambystoma mexicanum*;[81]

the rodenticide thiosemicarbazide, which produced abnormal digits
and limb articulation in *Rana sylvatica* tadpoles.[82]

Cowman and Mazanti[63] echo Berrill and his colleagues[83] in concluding that actual effects of pesticides on amphibians vary with species, age,
and environmental conditions. They continue this reasoning by pointing
out that no one particular species is more or less susceptible to pesticides,
so that no single species will serve as a bioindicator for environmental
contamination.

Metals also induce amphibian malformations and Linder and Grillitsch[84]
have summarized the literature. Metal compounds known to be teratogenic include cadmium chloride (*Xenopus laevis*[85]), potassium dichromate
(*Rana tigrina*[86]), lead nitrate (*Bufo arenarum*[87]), and mercuric chloride (*Bufo
fowleri, B. punctatus, Gastrophryne carolinensis, Hyla chrysoscelis, Rana grylio,
R. pipiens*[88]). Linder and Grillitsch also show that heavy metal concentrations, and consequently their toxicity, are highly dependent on pH.

Organic contaminants also have teratogenic effects on amphibians, and
Sparling[89] has summarized this literature. Sublethal levels of polychlorinated biphenyls (PCBs) produce skeletal defects, including scoliosis and
abdominal swelling.[90] DDT produces abnormal jaws.[91] *Rana catesbeiana*
tadpoles exposed to "Bunker C, number 6 fuel oil" developed edema.[92]
Sparling points out that amphibians may develop tolerance to organic
contaminant exposure.

Chromosomal damage A much overlooked paper assembled by Les
Lowcock and his colleagues[72] demonstrates abnormal DNA profiles from
Rana clamitans in populations that exhibit high rates of malformations (as
shown by Ouellet and his colleagues[93]). Lowcock and his coworkers used
a method called flow cytometry and measured something called half-peak
coefficients of variation, as well as variation in genome size. They found
that malformed juveniles had significantly higher half-peak coefficients
of variation, and therefore higher rates of genomic disruption, than normal individuals. They write (p. 241): "The different classes of DNA damage found in this study are reflective of either acute or cumulative pesticide

toxicity . . ." A similar study using the same techniques on the same species and *Rana palustris* by Bly and his colleagues[94] in southeastern Minnesota, where malformations are not a problem (2 of 796 animals were malformed, a rate of 0.25%), showed no evidence of genetic damage.

PARASITES

A careful reading of the literature shows that "the parasite hypothesis" is in fact two hypotheses that have evolved one into the other. Portions of the latest version of the parasite hypothesis have been well worked out and are unquestioned; other portions are not well understood at all. The notion that parasites cause amphibian malformations was first presented by Stan Sessions and Stephen Ruth in 1990.[17] Working in California, they noted that a parasite they tentatively identified as *Manodistomum* sp. infects both Pacific treefrogs (*Pseudacris regilla*) and long-toed salamanders (*Ambystoma macrodactylum*) to produce multiple limbs or multiple limb segments, with the duplicated portions oriented in a mirror-image fashion. (If this sounds confusing, hold your hands out in front of you, palms down. In this position your hands are mirror-image oriented. Now imagine your forearms arise from a common elbow. These are the types of malformations [although occurring predominantly in the hindlimbs] Sessions and Ruth observed.)

Sessions and Ruth also noted the position of the encysted metacercariae of this parasite, and proposed that these metacercariae mechanically disrupt the apical ridge ectoderm along the leading edge of the developing limb bud. When apical ridge ectoderm is divided, each portion reorganizes to form a complete and appropriate limb segment. In their words parasites: ". . . may simply act as solid obstacles that first injure limb tissues and then passively disrupt morphogenic processes during subsequent wound healing and continued limb development" (p. 38). The theory behind this idea was presented a decade earlier by Sue Bryant and her colleagues.[95]

Sessions and Ruth then brought their system into the laboratory. In the lab they inserted glass/resinous beads (a mimic of an encysted metacercaria) into developing limb buds and, *at the point where the beads were inserted*, achieved some of the mirror-image duplications they observed in wild-caught animals.

Finally, Sessions and Ruth noted that frogs and salamanders with too many limbs have difficulty moving, and hypothesized that infected animals are more vulnerable to snakes, which are one type of amphibian predator and the primary host for trematode parasites, tentatively identified as *Manodistomum* sp. (see also Sutherland[96]). In essence, by creating amphibian limb malformations, these trematodes promote predation by snakes and facilitate the completion of their life cycle. Extremely elegant, as are the life histories of many parasites.

The first parasite paradox   Problems for the Sessions and Ruth parasite hypothesis arose about five years after its publication, when the malformed amphibian phenomenon first arose and people sought explanations. While parasites were widely considered a likely cause, not all malformations look the same (Chapter 2), and most of the malformations observed in Minnesota were not the mirror-image duplications of Sessions and Ruth (the title of their paper was "Explanation for naturally occurring supernumerary limbs in amphibians"). Instead the Minnesota malformations were predominantly missing limbs and limb segments. Sessions and Ruth cited failed predation as the likely cause of missing limbs (i.e., duplicated limbs were due to parasite infections, missing limbs to failed predation), but the evidence for a general application of this idea was missing or contrary to field and lab observations (see sections on Predation and Landscape, pp. 134 and 121, respectively).

These observations cast severe doubt on the Sessions and Ruth parasite hypothesis as applied to the malformed frog phenomenon. And these observations also produced the first parasite paradox—while the Sessions and Ruth hypothesis was accepted by developmental biologists as sound science, it did not account for most malformation types observed by field biologists.

The second parasite paradox   In the late 1990s, Sessions and Ruth's work was re-examined by Pieter Johnson, Kevin Lunde, and their colleagues.[34] The first thing they realized was that key facts of the Sessions and Ruth hypothesis were incorrect; the parasites that caused malformations were not *Manodistomum*, but rather *Ribeiroia ondatrae*. This meant that the primary hosts (and thus the amphibian predators) were not snakes, but birds (and, as is now known, mammals also[36]).[97] While Sessions and Ruth erred in both the identification of the parasite and the primary host, they cannot be taken to task too greatly for this. They sent their material to a professional parasitologist; apparently two species were present and the parasitologist identified the wrong one.

But just as importantly, Johnson and his colleagues[34] show (their table 1, p. 803; reproduced in part here as Table 5.1) that *Ribeiroia* induce a wide range of hindlimb malformations in Pacific treefrogs (*Pseudacris regilla*), but not forelimb, cephalic, or axial malformations, which in their study comprised a little under 5% of the observed malformations in the field. They note (p. 804): "The types of abnormalities produced in this experiment encompass many of the abnormalities described in reports from across the continent."

This new hypothesis of Johnson and his colleagues—*Ribeiroia* infection followed by the induction of many malformation types, followed by predation by birds (Fig. 5.8)—is the "parasite hypothesis" referred to today, although just about every publication on the issue tends to lump Sessions and Ruth's hypothesis with the one proposed by Johnson and his colleagues.

While Johnson's work clarified Sessions and Ruth's premise, it also introduced a new problem; but first, some background. The mechanism Sessions and Ruth proposed for limb duplication—mechanical disruption (see also the paper by Stopper and his colleagues[11])—makes sense from a developmental biologist's perspective. In fact, one of the first experimental manipulations developmental biologists require undergraduates to accomplish is to divide apical ridge ectoderm to produce multiple limb segments. If students cannot do this, they are not offered bench space in the lab.[98] But Johnson and his colleagues noted that *Ribeiroia* metacercariae do not tend to occur in distal portions of the limb where many malformations occur (see radiographs in Chapter 2, especially Fig. 2.27), but rather near the body in the pelvic region. In particular, Johnson and his colleagues dissected wild-caught animals (p. 802) and write that *Ribeiroia* "metacercariae were highly localized in the tissue around the pelvic girdle and hindlimbs, often in close association with abnormal or extra limbs." And in their 2002 paper, Johnson and his colleagues write (p. 162):[20] "Within infected anurans, the parasite exhibited a nonrandom distribution, with the majority of metacercariae embedded around the base of the limbs and the tail resorption area. Correspondingly, malformations associated with *Ribeiroia* infection typically involved the limbs,

FIGURE 5.8   The life cycle of *Ribeiroia ondatrae*. From Johnson and Lunde.[97] Used with permission of the University of California Press.

including extra, missing, and malformed fore- and hind limbs . . ." Stopper, Sessions and their colleagues describe how this pattern of localization happens (p. 254):[11]

> We observed that the trematode cercariae are released from the snail and, once they contact a tadpole, most of them actively target the hind limb bud regions, especially preferring folds and indentations around the base of the limb buds and tail. Rarely, a cercaria was observed crawling into the mouth, spiracle, or cloaca. Multiple cercariae form cysts on the surface of the skin and penetrate into the tissues in and around the limb buds over the next few days.

In fact, the method of transmission of all tadpole-infected trematodes is similar.[99] Trematodes infect tadpoles by first attaching themselves to the tadpoles' skin and then crawling to the point of entry.[100] Tadpoles that are able to dislodge cercariae by burst swimming and by brushing them off may reduce their infection rates.[101] Taylor and colleagues disagree with Stopper, Sessions and their colleagues that trematode cercariae "actively target the hind limb bud regions." Instead, Taylor and colleagues observe (using high-speed videography) that the "dead-water zone" behind the torso and next to the base of the tail provides a place where cercariae can lodge without being shaken or brushed off. Taylor's group then notes that this time window is short, because (p. 703):[101]

> Once the distal elements of the hind limbs have differentiated, the hind limbs become motile. From then on, intrinsic limb movements are possible and can help shake cercariae out of the dead-water zone. Indeed, metacercarial cysts are rarely, if ever, observed in the hind limb distal to the femurs.

Because *Ribeiroia* tend to cluster in the inguinal region, and parasites are only known to produce malformations through direct mechanical disruption of limb bud tissues, the current parasite hypothesis cannot explain a variety of amphibian malformations, including forelimb, distal hindlimb, cranial, pigment, and systemic malformations. This, then, is the second parasite paradox: Whereas Johnson's hypothesis accounts for many (but not all) malformations observed in Pacific treefrogs by field biologists (see Table 5.1), the known mechanism through which parasites affect limb development cannot explain most amphibian malformation types.

Searching for a mechanism   Developmental biologists challenged the parasite advocates with the question: How can a parasite metacercaria located in the groin produce a malformed foot without affecting any intermediate structures (e.g., Fig. 2.27)? There has been speculation. In 1990, Sessions and Ruth note (p. 45):[17]

> The results of the bead implantation experiment indicate that mechanical disruption of developing limb tissues by inert objects similar in size and shape to trematode cysts is sufficient to stimulate the outgrowth of

TABLE 5.1. *Malformations in Pacific Treefrog Tadpoles*

Malformation Type	Light Ribeiroia (%)	Intermediate Ribeiroia (%)	Heavy Ribeiroia (%)	Field (%)
Cephalic and axial				
Missing eye	0	0	0	0.5
Abnormal jaw	0	0	0	0.6
Open wound	0	0	0	1.1
Other	0	0	0	0.4
Forelimb				
Missing limb	0	0	0	1.1
Missing digit	0	0	0	0.7
Cutaneous fusion	0	0	0	0.2
Extra digit	0	0	0	0.1
Extra limb	0	0	0	0
Other	0	0	0	0.1
Hindlimb				
Missing limb	11.8	18.6	25.5	6.5
Missing digit	0	0	2.1	4.2
Cutaneous fusion	17.6	9.3	12.8	5.9
Bony triangle	5.9	4.7	0	3.4
Extra digit	2.9	6.9	0	4.8
Extra limb	32.3	44.2	55.3	50.7
Femoral projection	11.7	2.3	2.1	12.2
Other	17.6	13.9	2.1	7.6
# Malformations/ animal	1.7	2.5	2.9	1.38

Note: Malformations produced by experimentally infecting Pacific treefrog (*Pseudacris regilla*: Hylidae) tadpoles with *Ribeiroia* cercariae at three densities (light, intermediate, and heavy) with a comparison of malformation types collected in the field (table 1, p. 803 of Johnson and his colleagues[34]). These malformation rates are similar to those found in Pacific treefrogs in the field by Sessions and Ruth[17] (their table 1, p. 39).

supernumerary limb structures. However, *these results do not preclude the possibility that the metacercaria cysts also secrete a growth-stimulating substance* [author's italics], but they do show that such a scenario is not necessary to explain the observed abnormalities.

But in 1999, Sessions and his colleagues[102] disregard the chemical se-
cretion hypothesis and instead compare the malformation types induced
by retinoids only with malformation types induced by *mechanical* disturbance by trematode parasites (p. 801):

> The effects of retinoids and *mechanical perturbation* [author's italics] on
> amphibian limbs allow predictions of specific kinds of morphological
> abnormalities. By comparing the morphological patterns found in multi-
> legged amphibians from natural populations with these predicted patterns,
> it should be possible to identify the most parsimonious explanation for this
> kind of deformity.

In 2002, Stopper, Sessions, and their colleagues[11] continue to explore
mechanical disruption as the sole cause of limb malformations (p. 258):

> These results indicate that the probable mechanism by which trematode
> cyst infestation causes limb deformities in frogs is *perturbation of the spatial
> organization of cells* [author's italics] in the developing limb buds followed by
> intercalation.

Considering Sessions and his colleagues' work on malformations as a
whole, in 1990, parasitic cysts were said to cause mechanical disturbance
but chemical effects remained in play, while in 1999 and 2002 chemical
effects are all but excluded. It must be noted that the emphasis in each
of these publications is on multilegged animals (see titles of the 1990
and 1999 publications in endnotes 17 and 102, and note that all 14 photographs of malformed *Rana* in the 2002 paper[11] show duplications or
multiplications).

Johnson and his colleagues also weigh in (p. 804):[34] "The mechanism
through which *Ribeiroia* interferes with amphibian limb development remains unknown, but probably involves chemical or physical distur-
bances—acting independently or in concert—of the developing limb
bud." In support of this statement they state (their endnote 17): "Complex
host parasites can mimic both invertebrate and vertebrate host hormones,"
and cite two papers—a 1970 paper by Mueller and a 1990 paper by Helluy
and Holmes.[103] They continue: "Elucidation of the mechanism [of
Ribeiroia's effects] may offer new insights into limb development, especially

if the trematode produces a vertebrate growth factor mimic." But rather than address this question of mechanism directly, Johnson and his colleagues seek general acceptance for their hypothesis by showing, in a series of papers, that *Ribeiroia* infection correlates with the appearance of malformations in other species and from other parts of the United States.

Sessions and his colleagues' notion that parasites cause malformations by mechanical perturbations of developing limb tissue, while undoubtedly true, fails to explain most malformed frog types found in nature. The problem, again, is that in eastern North America, in *Bufo boreas* (work by Johnson and his colleagues;[30] Table 5.2), and in *Rana* (predominantly) *pipiens* (several authors; Table 5.3), multilegged animals form only a small subset of the malformation types exhibited in nature (Chapter 2). Therefore, the notion that mechanical disruption of limb buds by *Ribeiroia* metacercariae causes most amphibian malformations falls short as a general theory. (In fact, this statement is true about every hypothesis of causes of amphibian malformations.) Despite all conjecture, the mechanism by which *Ribeiroia* metacercariae affect distal structures, as well as other structures not directly targeted by cercariae, remains unknown.

More data, less confidence: Bufo boreas   In 2001, Johnson and his colleagues published experimental results in western toads (*Bufo boreas*), comparing the relation between malformations produced by *Ribeiroia* infection in the laboratory and malformations found in nature (their table 1 [p. 372] is summarized here as Table 5.2).[30] In presenting these results, Johnson and his colleagues point out that, experimentally, *Ribeiroia* induced severe limb malformations in animals that survived to metamorphosis. Further, the frequency and severity of malformations increased with *Ribeiroia* exposure while survivorship declined. They note that all malformations induced involved the limbs, and that there was a diversity of malformation types. Cutaneous fusions were the most common malformation induced, while polymely and taumely ranked second and third, respectively (Table 5.2). In the field, even though *Ribeiroia* were present at all sites, the pattern of malformations differed from the experimental results; hindlimb ectrodactyly, hemimely, and apody predominated (Table 5.2). This was in clear contrast to the experimental data, where hindlimb ectrodactyly, hemimely, and apody accounted for only about 3% of the induced malformations. The percent similarity in malformation types between laboratory induced

TABLE 5.2. *Malformations in Western Toad Tadpoles*

Malformation Type	Light Ribeiroia (%)	Intermediate Ribeiroia (%)	Heavy Ribeiroia (%)	Field (%)
Cephalic and axial				
Missing eye	0	0	0	2.1
Abnormal jaw	0	0	0	0.7
Open wound	0	0	0	2.8
Other	0	0	0	0.7
Forelimb				
Missing limb	8.7	8.2	0	3.5
Missing digit	0	1.6	0	6.9
Extra digit	0	0	0	0
Extra limb	0	3.3	0	0
Other	0	0	0	1.4
Hindlimb				
Missing limb	0	1.6	0	*23.7*
Missing foot	0	0	2.2	*9.0*
Missing digit	0	3.3	2.2	*23.6*
Cutaneous fusion	**34.8**	**18.0**	**28.9**	3.5
Bony triangle	**8.7**	**14.8**	**17.8**	4.2
Extra digit	**13.0**	**14.8**	**11.1**	2.8
Extra foot	0	4.9	4.4	0
Extra limb	**17.4**	**14.8**	**15.6**	3.5
Femoral projection	0	3.3	4.4	0.7
Micromely	8.7	1.6	0	2.8
Hyperextension	0	1.6	2.2	0
Other	8.7	8.2	11.1	8.3
# Malformations/ animal	1.8	2.4	2.5	1.1

Note: Malformations produced by experimentally infecting western toad (*Bufo boreas*: Bufonidae) tadpoles with *Ribeiroia* cercariae at three densities (light, intermediate, and heavy) with a comparison of malformation types collected in the field (table 1, p. 372, of Johnson and his colleagues[30]). Note the high rates of missing hindlimb elements in the field data (italic text) compared with the experimental data, and the high rates of cutaneous fusions, bony triangles, and extra hindlimb elements in experimental data (bold text) compared with field data.

and field collected animals was only about 30%—far short of a general explanation. Of the seven field sites examined, only one site exhibited a pattern of malformations closely aligned to the experimental data, with a percent similarity of 72.4%.

In this western toad dataset, the congruence between experimental data and field data seen in the Pacific treefrog dataset is notably absent (compare Table 5.2 with Table 5.1). In particular, observe the high rates of missing hindlimb elements in the field data, and the high rates of cutaneous fusions, bony triangles, and extra hindlimb elements in the experimental data. Johnson and his colleagues recognize the discrepancy between the responses of Pacific treefrogs and western toads, and write (p. 374):[30]

> Collectively, these differences in malformation types and their relative frequencies between Pacific treefrogs and western toads indicate that amphibian species may respond differently to identical numbers of *R. ondatrae* cercariae.
>
> Western toad larvae . . . demonstrated a weaker malformation response to *R. ondatrae* infection relative to the response of Pacific treefrogs . . .

And "This disparity reflects either a reduced opportunity for *R. ondatrae* to utilize western toads as effective intermediate hosts or a reduced ability to do so."

In their followup (2002) paper, Johnson and his colleagues write (p. 163):[20] "The role of *Ribeiroia* in causing the morphological abnormalities observed in the remaining . . . amphibian species is more ambiguous. The abnormalities of *Bufo boreas* from aquatic systems with *Ribeiroia* showed a greater percentage similarity with the compiled abnormalities in anurans from sites without *Ribeiroia* than with the abnormalities of *B. boreas* larvae infected with *Ribeiroia* in the laboratory." In other words, malformations induced in the laboratory were different from malformations found in the field, and the lab versus field distinction was a better predictor of malformation type than was the presence of *Ribeiroia* in the field.

Johnson and his colleagues attempt to explain the differences in response between Pacific treefrog and western toads as follows (pp. 374, 376):[30]

> The difference in sensitivity may be due to several behavioral or physical distinctions between the two anurans. Western toad larvae tend to be more

active than those of Pacific treefrogs. During the experimental exposures, toad larvae swam vigorously for much of the 120-min period, potentially reducing the ability of *R. ondatrae* cercariae to attach and penetrate the skin. In contrast, treefrog larvae were more sedate, swimming only sporadically. . .

Toads and treefrogs also differ in skin properties. Toad epithelium contains noxious bufodienolides and is therefore unpalatable to many would-be predators. These toxins are particularly effective against birds, fishes, and mammals, with mixed results for amphibian and invertebrate predators. An amphibian host that is unpalatable to birds or mammals could reduce or halt *R. ondatrae* transmission between its second intermediate (amphibian or fish) and definitive hosts (birds and mammals), thereby selectively favoring infection of an alternate amphibian host species.

It is interesting to read these speculations, keeping Gould and Lewontin's critique of the adaptationist programme in mind. Six of the seven sites Johnson and his colleagues sampled did not follow the *Pseudacris regilla* model, but interestingly, one did (72.4% similarity). If the explanations for these data hold (toad activity and unpalatibility), why were toads from one site not mobile or unpalatable and thus susceptible to *Ribeiroia* infection? Why not consider an alternate interpretation: one *B. boreas* site follows the *P. regilla* model, while the other sites offer a second model, perhaps parasites in combination with some other factor, perhaps not parasites at all? Such hypotheses could generalize our knowledge of causes.

This rationalization of Johnson and his colleagues also falls apart when we examine *Ribeiroia* infection data from other *Bufo* species. It is clear from the data on *Bufo americanus* (which also have bufodienolides) from the CTG site in Minnesota (see Table 5.4, below) that there is little resistance to *Ribeiroia* infection by bufonids. An alternate interpretation, one that is independent of adaptationist speculation but fits the data, is one that Johnson and I used in our 2003 paper[33]—that bufonid tadpoles, being small, represent a more difficult target for *Ribeiroia* cercariae. This interpretation may not be correct, but at least it is worth testing. A second alternate hypothesis, proposed by Taylor and his colleagues,[101] is that, because of their toxicity, bufonids have no need to be cryptic. Unlike ranids, bufonids can effect the violent movements needed to shake off metacercariae before they can encyst, without concern for whether or not they attract predators.

Even more data, even less confidence: Rana pipiens   In their 2001 paper on western toads, Johnson and his colleagues write (p. 377):[30] "In future experiments, the specific effects of *R. ondatrae* on various ranids need to be examined, as species of this group have been frequently reported with limb malformations in the Midwestern U.S.A. and Canada." In fact, this work has now been done.

In 2002, Stopper, Sessions, and their colleagues[11] experimentally induced malformations in northern leopard frogs and wood frogs by two mechanisms: exposure to *Ribeiroia* cercariae and experimental limb bud rotations. They found that both treatments produced a range of polymelies (extra limbs or limb segments), including anterior, posterior, and dorso-ventral mirror-image duplications, mirror-image triplications, and proximo-distal duplications (their table 1, p. 254, reproduced in part here as Table 5.3). Other malformations induced, but not included in the presentation of their results, were "slightly abnormal bone structures," small or incomplete structures, and skin fusions. In 2003, Anna Schotthoefer led a team of biologists at the National Wildlife Health Center, in Madison, Wisconsin, that also examined the effects of *Ribeiroia* metacercariae on developing northern leopard frogs.[19] Both of these studies report the primary malformation type as polymely (limb duplications and triplications in Stopper and Sessions' group [their table 1]; multiple limbs, multiple digits, and multiple phalanges on digits in Schotthoefer's group [their table 2; p. 1149]). As an aside, while Stopper and Sessions' group concludes (p. 261), ". . . the ability of cysts to induce the *entire range* [italics mine] of deformities found in the wild, speak[s] to the great role that trematode cysts may be playing," Schotthoefer's group (p. 1151) writes "Notably absent from the malformations observed in *R. pipiens* in this study were missing or partially missing limbs . . ."

There have been more field data collected on malformations in ranids, especially *Rana pipiens*, than any other group. Studies include Ouellet and colleagues,[93] Helgen and colleagues,[15] Converse and colleagues,[104] Meteyer and colleagues,[5] Lannoo and colleagues,[33] and Levey.[44] Data compatible with the *Ribeiroia* experimental studies are presented in papers by Meteyer and her colleagues, Converse and her colleagues, and by Levey, and were used in conjunction with the experimental data of Schotthoefer's and Stopper and Sessions' groups to build a table for *Rana* similar to the tables Johnson and his colleagues constructed for *Pseudacris regilla* and *Bufo boreas* (Table 5.3).

The first notable feature of Table 5.3 is the large number of cephalic/axial and forelimb malformations found in *Rana*. These malformations constituted 20.2% of the malformations found by Levey and 13.5% of the malformations found by Converse and colleagues (Meteyer and her colleagues did not report on these malformation types, instead choosing to save their data for a subsequent paper that, to date, has not been published). These percentages exceed by about a factor of 10 the "normal levels of trauma, predation, and developmental error" suggested by Johnson and his colleagues (p. 163)[20] to explain non-*Ribeiroia* malformations in *Bufo boreas*.

The second notable feature of Table 5.3 is the large number of missing hindlimbs and hindlimb segments found in field-collected animals compared to *Ribeiroia*-exposed animals. All field workers examining *Rana*, and midwestern/eastern species in general, note that missing limbs and limb segments constitute the most common malformation encountered.

The third notable feature of Table 5.3 is the high rate of multiple limbs or limb segments induced by *Ribeiroia*. Stopper, Sessions, and colleagues[11] also report a high percentage of bony triangles induced by *Ribeiroia*, about half the rate Meteyer and her colleagues[5] found in field-collected animals.

A comparison of the hindlimb malformations induced experimentally by *Ribeiroia* and the malformations found in field populations for *Pseudacris regilla* (from Table 5.1), *Bufo boreas* (from Table 5.2) and *Rana* (predominantly *pipiens*; from Table 5.3) is given in Figure 5.9.

The data in Figure 5.9 suggest that the response of *Rana* is similar to the response seen in *Bufo boreas*. Furthermore, the persuasive congruence between laboratory data and field data used to argue for *Ribeiroia* causing malformations in *Pseudacris regilla* (the gold standard as far as these data sets are concerned), and for *Ribeiroia* causing malformations in general, is not only missing in *Bufo boreas*, as was recognized by Johnson and his colleagues (p. 374),[30] but is also missing in *Rana*. Further, because *Rana* are not toxic, and because the frequency of cephalic/axial and forelimb malformations are many times higher than background rates (Tables 5.2 and 5.3), the explanations Johnson and his colleagues give for the differences between *Bufo boreas* responses and *Pseudacris regilla* responses cannot hold for *Rana*.

TABLE 5.3. *Malformations in Ranid Frog Tadpoles*

Malformation Type	Schotthoefer et al.[19] Ribeiroia (%)[b]	Stopper et al.[11] Ribeiroia (%)[c]	Meteyer et al.[5] Field (%)	Levey[44] Field (%)	Converse et al.[104] Field (%)
Cephalic and axial					
Missing eye	0	0	—[d]	4.0	4.5
Abnormal jaw	0	0	—	0	1.8
Open wound	0	0	—	0.8	0
Other	0	0	—	0.8[e]	1.8[f]
Forelimb					
Missing limb	0	0	—[d]	0.8	3.6
Missing foot	0	0	—	4.0	0
Missing/short digit	0	0	—	9.0	?[g]
Cutaneous fusion	0	0	—	0	0
Extra digit	0	0	—	0	0
Extra limb	0	0	—	0	1.8
Other	0	0	—	0.8[h]	?[f]
Hindlimb					
Missing limb	0	0	7.0	4.9	12.7
Missing foot	0	0	35.0	28.6	1.8
Missing/short digit	2.8	0	24.2	33.5	?[g]
Cutaneous fusion[i]	8.3	?[j]	19.7	0	0
Bony triangle	0	**8.8**	2.5[k]	0.8[k]	0
Extra digit	**25**	0	2.5	0	0
Extra foot	0	0	3.2	0	1.8
Extra limb	**22.2**	**44**	5.1	0	0
Femoral projection	5.6	0	0	0	0
Micromely	0	0	1.3	4.0	0
Hyperextension	0	0	0		0
Other[l]	33.3	?[j]	8.3	3.2	?[f]
# Malformations/ animal	2.0	Not given	Not given	1.2	Not given

Additionally, it is worth noting what each of the workers examining ranids in the field observed about the presence or absence of *Ribeiroia*. Eaton and his colleagues, working in Alberta and Saskatchewan, write (p. 286):[37]

TABLE 5.3. *Malformations in Ranid Frog Tadpoles (continued)*

Note: Malformations produced by experimentally infecting ranid frog (predominantly *Rana pipiens*) tadpoles with *Ribeiroia* cercariae (Schotthoefer and colleagues [table 2],[19] and Stopper, Sessions, and colleagues [table 1][11]) are compared with *Rana* (predominantly *pipiens*) malformation types collected in the field (Meteyer and colleagues [table 2],[5] and Levey [table 9][44]). The far right column shows malformation types in a number of eastern U.S. species collected from U.S.F.W.S. refuges (Converse and colleagues [table 3][104]). The large number of missing hindlimbs and hindlimb segments found in field-collected animals are indicated by italic font. The high rates of bony triangles and multiple limbs or limb segments induced by *Ribeiroia* are indicated by bold font.

[a]Two other studies—Helgen and colleagues[15] and Ouellet and colleagues[93]—also address malformation frequencies in *Rana pipiens*. I did not include their data here because Helgen and her colleagues (table 4) combined forelimb and hindlimb malformations, and Ouellet and his colleagues combined shortened limb elements with shortened finger elements (tables 1 and 2). In general, though, the data in both papers are consistent with the field results of Meteyer and colleagues,[5] Converse and colleagues,[104] and Levey.[44]

[b]Percentages are from their table 2, p. 1149, and are taken from the "Total" column.

[c]Percentages are frequency of malformations produced by *Ribeiroia* cysts, the right column in their table 1 (p. 254). [d]Meteyer and her colleagues reported only on hindlimb malformations.

[e]Scoliosis.

[f]Of the malformations observed by Converse and her colleagues, 25.5% were classified as "unknown."

[g]Of the malformations observed by Converse and her colleagues, 32.7% were classified as "missing/deformed toes" without specifying forelimb or hindlimb.

[h]Emergence failure.

[i]Skin webbing, in the terminology of Schotthoefer and her colleagues.

[j]In their footnote to this table, the authors state: "Several specimens exhibit deformities that included slightly abnormal bone structures, hypomorphisms, and skin fusions; these are not included in the table."

[k]Rotation, in the terminology of Meteyer and colleagues, and Levey.

[l]Includes bone bridges and misshapen and reduced iliums.

A small number of the deformities we documented are consistent with descriptions and images of trematode-induced deformities. However, we found no evidence to link *Ribeiroia* with these deformities.

Helgen and her colleagues, working in Minnesota, write (p. 291):[18]

Most of the Le Sueur County malformed frogs had what appear to be parasitic cysts in the thigh muscles. The frequency of parasitic cysts in normal frogs is unknown because these frogs were released at the site. Two large adults collected from the Le Sueur County site had heavy loads of cysts but appeared normal in external and internal morphology. The Meeker

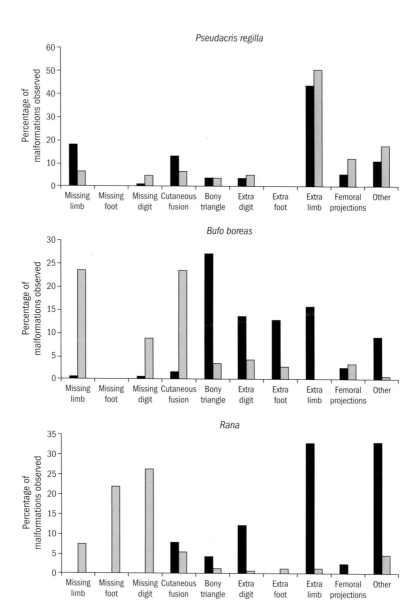

FIGURE 5.9　A comparison of hindlimb malformations induced experimentally by *Ribeiroia* (black bars) and malformations found in field populations (gray bars) for *Pseudacris regilla* (from Table 5.1), *Bufo boreas* (from Table 5.2), and *Rana* (predominantly *pipiens*; from Table 5.3). These data represent means; cells with "?" (Table 5.3), were not included in the analysis. Note the interspecific variation in these data, and that the tight congruence between *Ribeiroia* experimental and field data observed in *Pseudacris regilla* is missing in *Bufo boreas* and *Rana*.

County malformed frogs showed no visible appearance of parasite cysts in the legs.

Gardiner and Hoppe, also working in Minnesota, write (p. 213):[43] ". . . no trematode cysts were observed in association with either extra distal structures, hypomorphic limbs, or bony triangles."

Meteyer and her colleagues, working both in Minnesota and Vermont, write (p. 169):[5]

It should be emphasized again, however, that although polymelia was a predominant malformation in some limited circumscribed studies, they [sp.] were infrequent in the wide geographic area of our study and Canada and metacercariae were not found in the connective tissue of some of the malformed frogs.

Converse and her colleagues, working on federal lands in the Midwest and New England, write (p. 165):[104] "The trematode theory does not provide an answer for all abnormalities that have been observed." Ouellet and his colleagues, working in southern Quebec, write (p. 99):[93] "Inflammatory changes, parasites, and neoplastic alterations were not encountered in relation to limb structures."

Taylor and colleagues, working in Vermont, write:[39]

Examination of a subsample of individual specimens revealed no evidence of *Ribeiroia* infection. Similarly, representative samples of host snails from the sample wetlands did not demonstrate evidence of *Ribeiroia* infection.

Based on a report of samples sent to the U.S.G.S.-B.R.D. National Wildlife Heath Center, Rick Levey, also working in Vermont, writes (p. 2):[44]

No correlation has been found between incidence of abnormalities and parasite burden in newly-metamorphosed *R. pipiens*. Both normal and abnormal frogs can have heavy or light burdens of cysts.

Finally, Levey and members of David Skelly's lab write:[105] "*Ribeiroia* infection is not responsible for Vermont amphibian deformities."

Johnson and his colleagues have recognized this trend for interspecific differences in *Ribeiroia* infection rates and for interspecific differences in malformation types. In 2001, they published a paper documenting malformations in four species of amphibians (Pacific treefrogs, California newts [*Taricha torosa*], western toads, and bullfrogs) from the western United States and noted (p. 348):[16] "Among species, however, we documented substantial variation in abnormality composition." And on the next page: "The polymelous western toad and bullfrog metamorphs are most likely the result of *Ribeiroia* infection, given the rarity of mutational events, the high frequency of polymelia caused by *Ribeiroia* in the laboratory, and the scarcity of other agents shown to cause anuran supernumerary limbs in the field. However, the missing limbs in these species and in the California newt larvae are less easily explained." In 2003, they write (p. 65):[2] "Although parasitism by trematodes is the likeliest explanation for most outbreaks of amphibian deformities, it is certainly not the only cause . . ." These conclusions would seem to be inconsistent, at least in emphasis if not in fact, with other statements such as:

> *Ribeiroia* is almost always found where deformed amphibians are present, whereas chemical pollutants are found much less frequently. What is more, the parasitic infection seems to have skyrocketed in recent years, possibly reaching epidemic levels (p. 64).[2]

> Limb deformities [constitute] an emerging parasitic disease in amphibians.[31]

> [amphibian limb malformations constitute an] "emerging helminthiasis" [and (p. 332)] "the recent outbreak of deformities [might be due to] exogenous agents (e.g., pesticides, nutrient run-off, introduced fishes) . . . interacting with *Ribeiroia*, resulting in elevated infection levels . . ."[106]

> If a spreading epidemic of *Ribeiroia* accounts for much, or even most, of the increase in frog deformities seen in recent years, what accounts for the epidemic? (p. 65)[2]

But re-examine Tables 5.2 and 5.3 and ask yourself, why consider only hindlimb malformations? Certainly, at rates of greater than 5%, cranial malformations must be important, as are forelimb malformations (at 5–15%); forelimbs develop in gill cavities and are not likely to be affected by *Ribeiroia* cysts that penetrate the skin, as Stopper, Sessions, and their colleagues describe.[11] Somehow, hindlimb malformations have become the "type" malformation—other various malformations are either dismissed or ignored—and because of this, scenarios of causes that explain hindlimb malformations, and only hindlimb malformations (or not, see Figure 5.9), are seen to be plausible.

In summary, the data collected to date suggest that the current parasite hypothesis is less relevant the farther you get anatomically from proximate hindlimbs, the farther you get phyletically from *Pseudacris regilla* (an exception is the salamander *Ambystoma macrodactylum* [and there will always be exceptions]), and the farther you get geographically from the western United States.

Minnesota parasite data   The bulk of the lab/field evidence for and against the various forms of the current hypothesis that trematode parasites cause amphibian malformations is presented above. The original sources are cited (via quotes and endnotes) and both the data and the interpretations of the data are documented. While some researchers may find fault with certain details or emphases, by going to original sources and using original wording, this interpretation cannot be very far off.

But these are not the only data and interpretations available. I have access to a second dataset, on Minnesota malformed frog hotspots, that was published, but in a poorly circulated and poorly produced symposium proceedings.[33] Despite the fact that this work is rarely cited, it contains some of the most relevant data on the relation between parasite infections and malformations. Minnesota began the malformed frog phenomenon, and it burned the hottest there. Minnesota is situated in between the western states, where generalizations about malformations have been made, and the eastern states, where different generalizations about malformations have been made; it can be considered a battleground. No other state issued bottled water in response to the malformed frog phenomenon. No other state formed the basis for a book that was on the New York Times summer reading list.[4] No other state had the only classically trained parasitologist actively working on this problem systematically explore its malformed frog hotspots. And I was there, right next to Dan Sutherland, getting splattered with blood and frog goop, when he did his necropsies.

I have taped to my office wall a spreadsheet nearly two feet wide and five feet long representing the parasite data from the 2001 sampling trip

Dan Sutherland and I took to sample the hottest of the Minnesota hotspots (Chapter 3). The raw data from 286 necropsies of newly metamorphosed frogs are listed. Across the top of this spreadsheet are the various parasite groups: trematodes (flatworms), nematodes (roundworms), cestodes (tapeworms), and miscellaneous groups. Within the trematodes (obviously, the group that interests us the most at this time) are columns headed by *Ribeiroia ondatrae, Fibricola cratera,* "globbies" (a grouping of metacercariae that all possess globular excretory bladders and includes the species *Glypthelmins quieta, Auridistomum chelydrae* and undetermined metacercariae from several species), *Manodistomum* sp. (which represents the genus Sessions and Ruth originally thought was the cause of malformations in *Pseudacris regilla* and *Ambystoma macrodactylum*), kidney echinostomes (used by Thiemann and Wassersug, and others[101] to assess interactions between predators and parasite infection rates), and gill echinostomes (whose metacercariae encyst along the peripheral portions of cranial nerves exiting the base of the skull).

The data from the CTG site (Fig. 3.9) seem to me to best represent the mental image people have when they read papers describing the role of parasites in inducing amphibian malformations (Table 5.4). Each of the 26 American toads is infected with *Ribeiroia,* and some of the animals are heavily infected (30 or more *Ribeiroia* in animals of a centimeter and a half body length represents a heavy infection). In this table, gill and kidney echinostomes are combined and represent a smattering of incidental infection. As an aside, and as mentioned above, the heavy infections of these toads by *Ribeiroia* negates the speculation by Johnson and his colleagues that *Ribeiroia* avoid infecting bufonids because of the presence of bufodienolides.

The data on trematode parasite loads in newly metamorphosed northern leopard frogs (*Rana pipiens*) from sister wetlands, NEY (Fig. 3.2) and BUR (Fig. 3.8), in south-central Minnesota, better represent the sort of data we collected (Table 5.5). Note the large number of trematode groups represented. Despite being in close proximity, and despite sampling the same species, the parasite fauna from these two wetlands was markedly different. In our sample, the NEY site had more leopard frogs and/or a higher percentage of leopard frogs infected by *Clinostomum* sp. and globbies, while the BUR site had more leopard frogs and/or a higher frequency of leopard frogs infected by *Fibricola* sp., *Manodistomum* sp., kidney

TABLE 5.4. *Trematode Parasite Loads in Newly Metamorphosed American Toads*

SVL (mm)	Sex	Parasites				
		Clinostomum	Echino-stomata	Fibricola	Manodi-stomum	Ribeiroia
19	M		1			19
12	F					17
20	M					32
19	M					11
18	M		2			22
18	M		1			10
14	M		2			11
15	M		3			7
17	M					6
16	M		2			24
19	M					24
18	F					28
20	F					27
19	F					19
19	M					5
18	M		7			14
17	M					31
16			8			11
24	M					8
17	F					19
22	M		8			17
25	F		4			20
2.4	F		4			16
25	F					16
22	M		12			17
	M					13

Note: Based on a spreadsheet representing raw data collected by Dan Sutherland on trematode parasite loads in newly metamorphosed American toads (*Bufo americanus*) from the CTG site (Fig. 3.9), near St. Paul. Note first the impressively high numbers of *Ribeiroia* found in these small animals. Note also the comparative lack of metacercariae from other trematode groups.

TABLE 5.5. *Trematode Parasite Loads in Newly Metamorphosed Northern Leopard Frogs*

Site	SVL (mm)	SEX	Clino-stomum	Fibricola	Globbie	Manodi-stomum	Kidney Echino-stomata	Ribeiroia
NEY	44	F	*2*	22	*17*		4	3
NEY		M			*33*		19	
NEY	46	M			*54*		15	1
NEY	44	F					32	
NEY	44	M	*6*	1	*65*		8	
NEY	59	M			*7*	10	11	1
NEY	49	F			*47*		7	6
NEY	33	F	*2*		*37*		17	
NEY	48	F			*18*			
NEY	58	M			*32*		30	
NEY	45	M		17	*26*	13	2	
BUR	30	F			*3*	**1**	**100**	**3**
BUR	36	F			*12*	**3**	**50**	
BUR	33	M			*2*		**50**	**6**
BUR	33	M		**9**	*12*	**1**	**30**	
BUR	37	M		**21**	*13*	**1**	**51**	**6**
BUR	34	M		**7**	*7*	**3**	**38**	**12**
BUR	36	M		**1**		**14**	**70**	**3**
BUR	37	M			*1*	**4**	**39**	**2**
BUR	33	M				**3**	**45**	**5**
BUR	35	M		**5**		**1**	**75**	
BUR		M		**24**		**6**	**57**	
BUR	32	M		**3**	*7*	**1**	**63**	

Note: Based on a spreadsheet representing raw data collected by Dan Sutherland on trematode parasite loads in newly metamorphosed northern leopard frogs (*Rana pipiens*) from sister wetlands, NEY (Fig. 3.2) and BUR (Fig. 3.8), in south-central Minnesota. We sampled these wetlands on the same day. Note that NEY frogs were bigger (NEY water was warmer); and that at the time of our sampling the NEY site had both more animals and a higher percentage of animals infected by *Clinostomum* sp., and Globbies (italic font), while the BUR site had more animals and/or a higher frequency of animals infected by *Fibricola* sp., *Manodistomum* sp., kidney Echinostomata, and *Ribeiroia ondatrae* (bold font). *Alaria* sp. were not found in animals from either site.

Echinostomata and *Ribeiroia ondatrae*. *Alaria* sp. (not shown) were not found in northern leopard frogs from either site.

Note that at both the BUR and NEY sites, some frogs were infected by *Ribeiroia*, others were not. Comparing BUR with NEY, a higher percentage of northern leopard frogs were infected with *Ribeiroia* (58% vs. 40%) and numbers of cercariae per infection were higher (5.89 vs. 2.75 cercariae/infected animal).

As a third example from our spreadsheet, I present trematode parasite loads in newly metamorphosed and adult mink frogs (*Rana septentrionalis*) and green frogs (*Rana clamitans*) from sister wetlands CWB (Fig. 3.5) and MHL, in north-central Minnesota (Table 5.6). Despite their close proximity, in our sample the CWB site had both more frogs and a higher percentage of frogs infected by most trematode groups, including *Alaria* sp., kidney Echinostomata, and *Ribeiroia ondatrae*, compared to the MHL site. Within MHL, *Clinostomum* and *Ribeiroia* infections were restricted to mink frogs, while *Fibricola* and *Manodistomum* infections were restricted to green frogs, although these patterns are undoubtedly influenced by low infection rates, and thus small sample sizes.

As a fourth and final example from our spreadsheet, I show between-species differences in trematode parasite loads in newly metamorphosed and adult northern leopard frogs (*Rana pipiens*) and mink frogs (*Rana septentrionalis*) from the BLO site (Fig. 3.11), in northwestern Minnesota (Table 5.7). Both species were collected at the same time. At the time of our sampling, there were more individual northern leopard frogs infected, and a higher rate of infection per frog, by *Fibricola* sp. and *Manodistomum* sp. Mink frogs had more animals infected, and a higher rate of infection per frog, by *Alaria* sp. and kidney echinostomes. *Ribeiroia ondatrae* were not found in any animals of either species.

Several other trends emerged from the Minnesota parasite data. Encysted echinostome metacercariae were found in the kidneys of animals from every site. Other metacercariae, such as those from *Fibricola cratera* and ochetosomatids (*Manodistomum* sp.), were found in frogs from a majority of wetlands. In the remaining parasite species, there was strong site specificity; they could be present, often in high numbers, in amphibians from one site and completely absent from another. This was true of *Ribeiroia ondatrae*.

TABLE 5.6. *Trematode Parasite Loads in Newly Metamorphosed and Adult Mink and Green Frogs*

Species	Site	SVL	Sex	Alaria	Clino-stomum	Fibri-cola	Globbie	Manodi-stomum	Kidney Echino-stomata	Ribe-iroia
Rasep	CWB	39	F				2	1	*145*	*20*
Rasep	CWB	40	F				1		*425*	*27*
Rasep	CWB	42	F	*100*				1	*209*	*47*
Rasep	CWB	42	F	*44*			6		*199*	*14*
Rasep	CWB	44	M	*7*					*295*	*19*
Rasep	CWB	40	F	*46*			3		*129*	*23*
Rasep	CWB	40	M	*289*			3		*216*	*50*
Rasep	CWB	42	F	*78*					*205*	*52*
Rasep	CWB	41	F	*1*	1		2		*350*	*67*
Raclam	CWB	52	M				1		*275*	*12*
Rasep	MHL	38	F						*4*	
Rasep	MHL	36	M		**2**					
Rasep	MHL	37	M		**6**				*69*	**20**
Rasep	MHL	31	F							
Raclam	MHL	48	F				*1*			
Raclam	MHL	38	F							
Raclam	MHL	38	M							
Raclam	MHL	40	M				**8**		30	
Raclam	MHL	38	M				**8**			
Raclam	MHL	43	M				**5**		4	
Raclam	MHL	82	F			*19*				
Raclam	MHL	42	M						109	
Raclam	MHL	31	F					***1***	1	

Note: Based on a spreadsheet representing raw data collected by Dan Sutherland on trematode parasite loads in newly metamorphosed and adult mink frogs (*Rana septentrionalis*) and green frogs (*Rana clamitans*) from sister wetlands, CWB (Fig. 3.5) and MHL, in north-central Minnesota. Frogs are sorted first by sites (left hand column), then species within sites (second column from left). Despite their close proximity, the CWB site had both more frogs and a higher percentage of frogs infected by most trematodes, including *Alaria* sp., kidney Echinostomata, and *Ribeiroia ondatrae* (italic font) than the MHL site. Within MHL, *Clinostomum* and *Ribeiroia* infections were restricted to mink frogs (bold font), while *Fibricola* and *Manodistomum* infections were restricted to green frogs (italic bold font), although these patterns are undoubtedly influenced by low infection rates, and thus small sample sizes.

TABLE 5.7. *Trematode Parasite Loads in Newly Metamorphosed and Adult Northern Leopard and Mink Frogs*

Species	SVL	Sex	Alaria	Fibricola	Globbie	Manodi-stomum	Kidney Echino-stomata	Ribeiroia
Rapi	48	F		*3*	1	*22*	23	
Rapi	50	F		*2*	1	*45*	4	
Rapi	40	F				*22*		
Rapi	42	F		*18*		*52*	40	
Rapi	44	M				*65*	39	
Rapi	46	M				*9*	6	
Rapi	43	M		*2*	1	*12*		
Rapi	44	M		*2*		*20*	6	
Rapi	43					*50*	7	
Rapi	63	F		*1*		*331*	36	
Rasep	39	M		1		33	**3**	
Rasep	30	F	**5**				62	
Rasep	28	F	**80**				100	
Rasep	32	F	**9**				90	
Rasep	30	M	**19**			1	59	
Rasep	32	F	**10**	3		6	16	
Rasep	30	F	**6**				104	
Rasep	36	M				1	1	
Rasep	30	F	**244**				2	
Rasep	29	F	**77**					

Note: Based on a spreadsheet representing raw data collected by Dan Sutherland on trematode parasite loads in newly metamorphosed and adult northern leopard frogs (*Rana pipiens*) and mink frogs (*Rana septentrionalis*) from the BLO site, in northwestern Minnesota. Both species were collected at the same time; note that *R. pipiens* were larger. At the time of our sampling *Rana pipiens* had more animals and a higher percentage of animals infected by *Fibricola* sp. and *Manodistomum* sp. (italic font) while *R. septentrionalis* had more animals and a higher percentage of animals infected by *Alaria* sp. and kidney Echinostomata (bold font). *Ribeiroia ondatrae* were not found in any animals of either species.

As we note in our 2003 paper (summarized here as Fig. 5.10), heavy *Ribeiroia* infections were indicative of malformation hotspots (e.g., the CTG, CWB, and HIB sites), but lesser *Ribeiroia* infections might (e.g., the BUR, GEL, NEY, and ROI sites [Fig. 3.3]) or might not be associated with malformations (e.g., the MHL site; Fig. 5.10). Conversely, hotspots such as the CBA (Fig. 3.7), DOR, HYD, SUN, and TRD (Fig. 3.4) sites showed no evidence of the presence of *Ribeiroia*. Among the five reference sites, the two Iowa wetlands (OKB1 and OKB2), and the IWPA and BLO wetlands also showed no evidence of the presence of *Ribeiroia*.

Some *Ribeiroia* infections could be massive. In 1999, four severely malformed mink frogs necropsied from the CWB site harbored a mean intensity of 110 *Ribeiroia* metacercariae (range 96–125). In 2000, 12 northern leopard frogs (10 malformed) from the HIB site were infected with *Ribeiroia* (mean intensity 155.5; range 51–266). The only two apparently normal frogs in the 2000 HIB sample had the two smallest *Ribeiroia* infections (51 and 52 metacercariae, which still seem substantial). Malformations at the HIB site in 2000 included cutaneous fusions, truncations, bony protuberances, and soft tissue protuberances. In northern leopard frogs, it would not appear that parasitic cysts clear via an immune response.

Where *Ribeiroia* occurred, metacercariae were not found in every species of amphibian inhabiting the wetland. For example, at the MHL site, *Ribeiroia* metacercariae were found in mink frogs (n = 4), but not in green frogs (n = 9); in ROI *Ribeiroia* were found in leopard frogs (n = 11), but not in wood frogs (n = 4) or American toads (n = 1). This pattern may be due to sampling artifact. In general, species with longer larval stages have higher rates of *Ribeiroia* infection, and higher rates of infection among these species often correspond to higher numbers of amphibian species being infected. Our conclusions differ from those of Johnson and his colleagues,[20] who note (p. 156): "The presence of *Ribeiroia* at a site was associated with above-baseline frequencies (> 5%) of abnormalities among the amphibian species we examined . . . The mean number of *Ribeiroia* metacercariae per amphibian also was a significant predictor of the frequency of abnormalities."

At wetlands where *Ribeiroia* was present in every species of amphibian, infection rates varied across species. For example, at the CWB site, mink frogs (n = 9) averaged 35.4 *Ribeiroia* metacercariae per animal, while the single green frog sampled had 12, a lower number than any mink frog (again, this could be sampling artifact). Similarly, at the HIB site, wood frogs (n = 7) averaged 17.4 *Ribeiroia* metacercariae, northern leopard frogs (n = 11) averaged 7.9, and the one American toad had three. In our 2003 paper, we suggest that the low American toad infection rates at this site might be due to their small body size (and therefore small target area for roaming cercariae). This may or may not be true; American toads from the CTG site (17.1; n = 26) and wood frogs from the HIB site (17.4; n = 7), both small-bodied anurans, had high *Ribeiroia* infection rates, although infection rates must also depend on cercarial density.

Perhaps the most surprising finding is the distribution of amphibians infected with *Ribeiroia* metacercariae (Fig. 5.10B). Prior to this study, none of us had realized the strong tendency for *Ribeiroia* to occur predominantly in eastern Minnesota wetlands. This tendency extends to other sites in southeastern Minnesota and western Wisconsin (D. Sutherland, unpublished data). The majority of these sites are within the northern hardwood or mixed forested ecoregions.[107] In our sample, no *Ribeiroia* sites occurred within grassland regions, although I hesitate to generalize beyond our dataset.

In hotspots where *Ribeiroia* infections were absent, this parasite cannot be the cause of observed malformations. These sites include DOR, HYD, and TRD. In one of these sites (TRD), macrophyte beds that provide habitat for planorbid snail hosts were undeveloped at the time of our collecting; in three years of sampling this site, Hoppe (unpublished data) failed to find snails of any species. In the absence of host habitat, hosts, and metacercariae, it is difficult to argue for parasites as a cause of these malformations. Again, this conclusion is markedly different from the conclusions of Johnson and his colleagues,[20] who note (p. 162): "Among the sites we surveyed, the parasite *Ribeiroia ondatrae* was a powerful predictor of the presence and frequency of malformed amphibians in a population." And later on the same page: "The mean frequency of abnormalities in amphibians from sites supporting *Ribeiroia* was significantly greater than both the mean frequency in the absence of *Ribeiroia* and the expected baseline frequency."

The interest in *Ribeiroia*-induced malformations has triggered a series of papers on trematode-amphibian interactions. These papers show, typically using echinostomatids, interspecific differences in tadpole response,[101] tadpole stage related effects,[19,108] tadpole density related ef-

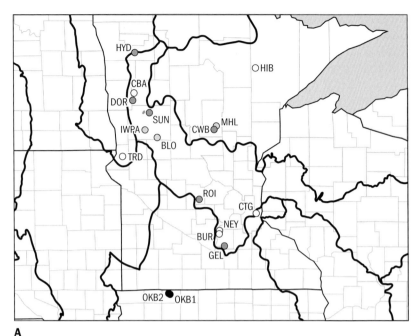

A

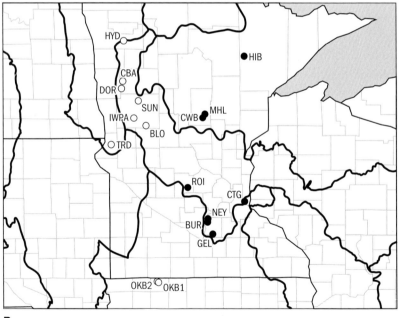

B

FIGURE 5.10 Opposite.   (A) A map of Minnesota and portions of surrounding states showing sites Sutherland, Kapfer, various colleagues, and I sampled during the late summer of 2001 (see also Fig. 3.1, which includes a key to sites). Dark-gray dots indicate natural wetlands considered to be hotspots; white dots indicate created wetlands considered to be hotspots; light-gray dots show Minnesota reference (control) sites—sites associated with wetlands; black dots represent two reference sites from northwestern Iowa—sites long studied by the author and, we felt, distant from impacts influencing Minnesota sites. Bold lines indicate ecoregion boundaries. (B) The same map, with sites where *Ribeiroia* metacercariae were found in necropsied frogs indicated by black dots, and sites with *Ribeiroia* metacercariae absent indicated by white dots. Note the strong regional specificity of these sites, and the independent distribution of *Ribeiroia* sites (black dots in B) and hotspots (dark-gray and white dots in A). Maps assembled by Robert Klaver.

fects,[109] trematode density related effects,[19, 110] and effects due to the type of wetland being sampled.[111] There is no "one size fits all" relationship between trematode infection and amphibian larvae.

Does the current parasite theory compromise a more general parasite theory? Looking at the aggregate parasite data, and knowing the complicated interactions between trematodes, tadpoles, and the environment, it's worth asking if the parasite hypothesis has been done a disservice by being simplified in an apparent effort to tell a consistent story. For example, by ignoring the fact that the only proven method by which *Ribeiroia* have an effect—mechanical disruption of developing limb-bud tissues—we exclude an understanding of why parasite infections appear, in some cases, to be related to other types of malformations, such as systemic problems, forelimb malformations, cranial malformations (including eye displacements), and pigment malformations. Further, attempting to establish parasites as a general cause by correlating malformation sites with the presence of *Ribeiroia* also has problems. For one, as all scientists know, correlation is not the same as causation. We know that northern leopard frogs containing 50 or more *Ribeiroia* cysts can develop normally, and so it is possible that malformed animals can have low- to medium-grade *Ribeiroia* infections but their malformations are due to some other factor(s).

Also, the correlation between malformation types produced by laboratory infection of *Ribeiroia* and malformation types found in field-collected animals is astonishingly poor in *Bufo boreas* and *Rana pipiens* (see Tables 5.2, 5.3, and Figure 5.9). Using their *Bufo boreas* data, Johnson and his colleagues calculate a coefficient of similarity between *Ribeiroia*-induced and field-collected malformations of only about 30%. Given this, as mentioned above, it is surprising to read the conclusion of Stopper, Sessions, and their group (p. 261) that:[11] ". . . the ability of [*Ribeiroia*] cysts to induce the entire range of deformities found in the wild, speak[s] to the great role that trematode cysts may be playing." Meanwhile, Schotthoefer's group,[19] not being tempted to play the game of combining data across species after interspecific differences have been clearly established, writes (p. 1151): "Notably absent from the malformations observed in *R. pipiens* in this study were missing or partially missing limbs . . ."

As well, any correlation between malformations and *Ribeiroia* infection was bound to fail when international or global datasets were considered. In 2000, Martin Ouellet published a comprehensive analysis of the malformed frog literature,[27] and in 2003, Pieter Johnson and his colleagues published a comprehensive analysis of the *Ribeiroia* literature;[36] Johnson and colleagues reported on all known occurrences of this parasite, regardless of host. I sorted this literature according to continent (with the obvious exclusion of Antarctica). Note that reports of malformed frogs are strongly *bimodal*, with 78 papers from Europe and 68 papers from North America, but only 21 papers from Australia/New Zealand, 12 from Asia, seven from South America, and four from Africa (Fig. 5.11A). In contrast, note that reports of *Ribeiroia* infection are strongly *unimodal*, with 86 papers from North America, five papers from South America, four papers from Africa, and one paper from Europe (Fig. 5.11B). Comparing these two graphics, note that *Ribeiroia* are rarely (one report) found in Europe, and cannot be the cause of the large number of malformations reported there.

A QUICK SUMMARY

Q: *Do chemicals cause frog malformations?*
A: Yes, there is undisputable evidence.

Q: *Do chemicals cause all frog malformations?*
A: No, see discussion above.

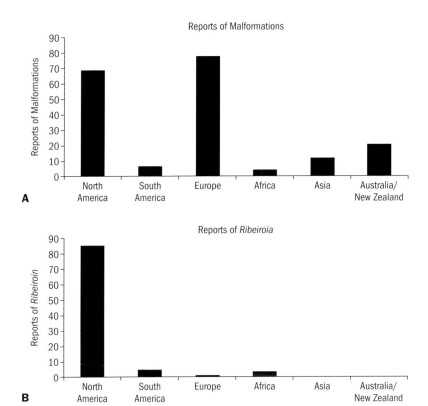

FIGURE 5.11 Reports of amphibian malformations (A) and *Ribeiroia* (B) by continent. Malformation data are taken from Ouellet;27 *Ribeiroia* data are from Johnson and colleagues36 (see text for details). Note that while the malformation reports are strongly bimodal, reports of *Ribeiroia* are strongly unimodal, with many fewer reports from non–North American sites.

Q: *Do parasites cause frog malformations?*

A: Yes, again there is undisputable evidence.

Q: *Do parasites cause all frog malformations?*

A: No. Problems in generalizing the parasite hypothesis include: (1) *Ribeiroia ondatrae* cysts are not found in malformed frogs from many sites; (2) mechanical disruption by parasitic cysts does a poor job of explaining all but proximal hindlimb malformations; (3) malformation types produced experimentally by *Ribeiroia* often do not correspond with malformation types found in the field; and (4) in places such as Alaska and the Old World, *Ribeiroia* are not known to exist in wetland ecosystems.

Q: *Are there other causes for malformations besides chemicals and parasites?*
A: Yes, and the lists given in Chapter 4 summarize this literature.

Q: *Does the presence of one cause negate other causes?*
A: There has been so much contention surrounding the discovery of causes of malformed frogs that finding one cause has tended to be interpreted as excluding other causes. This is a false assumption. For example, while CWB contains *Ribeiroia*, this wetland was also littered with frog and fish carcasses, so this trematode cannot be the only problem. *Ribeiroia* is not known to cause mass die-offs in normally-appearing tadpoles, adult frogs, or fishes.

Q: *Can the same cause produce more than one malformed frog type?*
A: Yes, for example see Tables 5.2 and 5.3, and Figure 5.9.

Q: *Can the same malformation type be produced by different causes?*
A: Yes, see Figure 2.36. Further, the polymelic animals from Ripley, Ohio, suggest a parasite cause, but Johnson and his colleagues have been unable to find either historical or recent evidence of *Ribeiroia ondatrae* from this site; the cause must be something else.

Q: *Can the same malformation type be produced in different species?*
A: Yes, compare Figures 2.1 and 2.2 (*R. pipiens*) with 2.3 (*R. catesbeiana*); Figure 2.9 (*R. pipiens*) with 2.10 (*R. septentrionalis*); Figure 2.13 (*R. pipiens*) with 2.14 (*A. crepitans*); Figure 2.35 (*P. regilla*) with 2.36 (*R. pipiens* and *R. sylvatica*); and the animals in Figure 2.29 (*R. pipiens* and *R. catesbeiana*).

Q: *Can the same malformation type be produced at different sites?*
A: Yes, compare Figure 2.1 (Minnesota-ROI) with 2.2 (Minnesota-WIN); Figure 2.13 (Wisconsin) with 2.14 (Indiana); Figure 2.35 (California) with 2.36 (Minnesota-NEY, TRD, CWB, and Alaska); and the animals in Figure 2.29 (Minnesota and California).

Q: *Are all malformations hindlimb malformations?*

A: No, and the rate of non-hindlimb malformations can be high (see Tables 5.2, 5.3).

Q: *If you solve the hindlimb malformation problem have you solved the amphibian malformation problem?*

A: No, as demonstrated in Chapter 2, the best description of the range of amphibian malformations includes all regions and most likely all systems of the body.

SIX

HUMAN MALFORMATIONS
AND CAUSES

The question that he frames in all but words
Is what to make of a diminished thing?

ROBERT FROST[1]

The fear that what was happening to frogs could happen to humans is the major reason why the malformed frog problem tipped in 1996. As with many ways of thinking about malformed frogs, the logic behind this concern is sound. Something in the water is causing frog malformations; developmental systems are conserved across vertebrates; therefore, if it's happening to frogs it could happen to us. Given this, it becomes useful to review human malformations (congenital anomalies) in light of frog malformations.

The rate of congenital anomalies in humans varies slightly from continent to continent and region to region, but globally it is remarkably constant, somewhere between 1.5 and 3.5% of live births.[2] This rate increases with age up to about five years, when other problems (e.g., heart defects, deafness, epilepsy, growth deficiencies, and mental deficiencies) not apparent at birth get noticed. Serious problems in the development of the brain, internal organs, and glands often have no adverse effect on prenatal growth but can cause serious postnatal secondary growth deficiencies.[2]

TYPES OF HUMAN MALFORMATIONS

This list of human congenital anomalies, along with their numeric codes, is known as the International Classifications of Diseases, 9th Revision,

Clinical Modification (ICD-9-CM Codes).[3] This system is used by hospital staff to standardize the reporting of birth defects.

XIV. Congenital Anomalies (ICD-9-CM Codes 740.0–759.9)

740   Anencephalus and similar anomalies

741   Spina bifida

742   Other congenital anomalies of the nervous system

743   Congenital anomalies of the eye(s)

744   Congenital anomalies of the ear(s), face, and neck

745   Bulbus cordis anomalies and anomalies of cardiac septal closure

746   Other congenital anomalies of heart

747   Other congenital anomalies of circulatory system

748   Congenital anomalies of respiratory system

749   Cleft palate

750   Other congenital anomalies of upper alimentary tract

752   Congenital anomalies of genital organs

753   Congenital anomalies of urinary system

754   Certain congenital musculoskeletal deformities

755   Other congenital anomalies of the limbs

756   Other congenital musculoskeletal anomalies

757   Congenital anomalies of the integument

758   Chromosomal anomalies

759   Other and unspecified congenital anomalies

Each of the list's whole number codes has subdivisions. For example, Code 755, Other congenital anomalies of limbs, is divided as follows:

755.0   Polydactyly

755.1   Syndactyly

755.2   Reduction deformities of the upper limb

755.3   Reduction deformities of the lower limb

755.4   Reduction deformities, unspecified limb

755.5   Other anomalies of upper limb, including the shoulder girdle

755.6   Other anomalies of lower limb, including the pelvic girdle

755.8   Other specified anomalies of an unspecified limb

755.9   Unspecified anomalies of an unspecified limb

The most common types of congenital anomalies are chromosomal abnormality syndromes (primarily ICD-9-CM code 758, but malformations encompassed in other codes can be produced by chromosomal abnormalities).[2] Chromosomal abnormality syndromes occur in about 5% of human pregnancies; many fetuses displaying chromosomal abnormalities are spontaneously aborted, and thus do not show up in statistics based on birth rates.

Triploidy occurs where three (rather than the expected two) copies of *each* chromosome are present (frequency of occurrence 1:100, with the majority of embryos spontaneously aborted).[4] Many chromosomal abnormalities consist of triplicate copies of *particular* chromosomes (humans normally have 23 pairs of chromosomes, each pair is labeled with a number) including: Trisomy 21 (Down Syndrome; frequency of occurrence 1:630–750 live births),[2] Trisomy 18 (Edwards syndrome or Trisomy E; frequency of occurrence 1:3,300–11,000 live births),[2] Trisomy 13 (Patau Syndrome or Trisomy D; frequency of occurrence 1:5,000 live births),[2] and Trisomy 8 (few reported cases).[2] Both Trisomy 21 and Trisomy 18 can occur in population clusters, and can co-occur in affected populations. Trisomies 2, 14, 15, 16, and 22 result in spontaneous abortions.

Additional chromosomal syndromes occur rarely and include 4p (Wolf-Hirschhorn) syndrome (frequency of occurrence 1:50,000 live births);[2] 5p (cri-du-chat or cry of the cat syndrome; frequency of occurrence 1:50,000 live births),[2] 13q (frequency of occurrence 1:12,000–22,000 live births),[2] 18p and 18q (frequency of occurrence 1:40,000 live births),[2] and 21q (perhaps 1:4,000 live births).[2] These syndromes are caused by chromosomal deletions, with "p" denoting short chromosomal arm deletions and "q" denoting long arm deletions.

Among the sex chromosomes, fragile X syndrome is caused by an expanded trinucleotide (CGG) repeat (frequency of occurrence 1:1000 male births);[2] XYY (Jacobs) syndrome (frequency of occurrence 1:1000 live births)[2] is caused by an extra copy of the Y, or male-determining, chromosome (frequency of occurrence 1:840 births);[2] XXY (Klinefelter) syndrome is caused by an extra copy of the X, or female-determining, chromosome, which is the most common cause of hypogonadism (small testes; frequency of occurrence 1:500 male births).[2] Females with the following chromosomal abnormalities have also been described, but cases are rare: XXX, XXXX, XXXXY, XXXXX, and XO (Turner syndrome;

frequency of occurrence 1:2,500 live births). Interestingly, no instances of similar duplications among Y chromosomes have been reported. Other types of chromosomal abnormalities leading to malformations have also been described.

A second grouping of human malformations is based on genetic mechanisms (rather than gross chromosomal aberrations, which admittedly are a form of genetic alteration, although they are not typically inherited), producing variously dwarfism, an elderly appearance (progeria), growth deficiencies, genital anomalies, neuromuscular findings, facial defects, and limb defects (see below).

CAUSES OF HUMAN MALFORMATIONS

Causes of human malformations are as variable as the types of malformations listed above. Known causes include:

1. Maternal and perhaps paternal age;[5]
2. Season of the year;[6]
3. Maternal diet, especially a vegetarian diet;[7]
4. Maternal nutrition, including abnormal levels of vitamin A intake;[8]
5. Maternal health,[9] including diabetes and Rubella;
6. Medical testing, especially chorionic villus sampling;[10]
7. In vitro fertilization;[11]
8. Prescription medications,[12] including ampicillin, oxytetracycline, anti-inflammatory medications, anti-nauseants, anti-epileptics, anti-microbials, asthma medications, and oral contraceptives;
9. Over-the-counter medications,[13] including bendectin, non-steroidal anti-inflammatory medications, and aspirin;
10. Illegal drug use,[14] especially cocaine;
11. Tobacco smoking/tobacco smoke;[15]
12. Alcohol consumption,[16] especially producing fetal alcohol syndrome;
13. Occupational exposure,[17] especially in the agricultural, leather, lumber, and health-care industries;
14. Air pollution;[18]

15. Water pollution/water additives;[19]
16. Environmental contamination;[20]
17. Pesticide exposure;[21]
18. Heavy metal exposure,[22] including lead, manganese, and arsenic;
19. Proximity to hazardous waste sites;[23]
20. Exposure to exogenous hormones/endocrine disruption/growth factors;[24]
21. Exposure to solvents or plasticizers.[25]

MISCARRIAGE RATES

In humans, early miscarriages are usually due to chromosomal division errors. In perhaps the best study on the subject conducted so far, the spontaneous abortion rate among fetuses producing human chorionic gonadotropin (and therefore living to at least post-conception day eight) is 31%.[26] Because some fetuses with severe defects never live to eight days and therefore never make chorionic gonadotropin, the spontaneous abortion rate must be higher, and some estimates (these are nothing more than educated guesses) are much higher. As an aside, among survivors of the atomic bomb blasts in Hiroshima and Nagasaki, there is no statistically demonstrable increase in major birth defects; in part because background rates of human birth defects are so high.

HUMAN MALFORMATION TYPES COMPARED WITH AMPHIBIAN MALFORMATION TYPES

Many human malformations typically result from chromosomally or genetically based developmental problems, and most frog malformations are typically environmentally induced developmental problems (but see Chapter 7); there is little overlap between types of human malformations and types of frog malformations. Even among similar types of malformations, humans and frogs show differences. For example, about 10% of humans with syndactyly (fusion of the fingers) hand exhibit an absence of pectoral (chest) muscles, the result of a primary defect in mesoderm. This suite of malformations is known as Poland anomalad, or Poland syndrome, named after the physician who first described this condition in 1841 (frequency of occurrence 1:17,000–32,000).[2] A similar suite of malformations has not been observed among amphibians.

A rare condition among humans termed limb reduction-ichthyosis,[2] or CHILD (congenital hemidysplasia with ichthyosiform erythroderma and limb defects) syndrome involves both limbs on the same side of the body. The clinical description is (p. 169):[2] moderate to severe shortening (hypomelia) of limbs, +/– absent scapula (shoulder blade) and diminished ribs, webbed elbow and knee. The skin on the same side of the body (and the demarcation between sides of the body is unusually distinct) is described as "thin, immature, slightly scaling . . ." (i.e., resembling fish skin, giving rise to the term ichthyosis). This unusual condition (< 50 cases known, including two siblings) is associated with the "partial to complete absence of cardiac septum with valve defects." Malformed frogs exhibit shortening of limbs, webbed joints, and skin (pigment) abnormalities, but rarely does this condition affect both limbs on the same side of the animal and involve pigment patterns on the thorax or abdomen.

A second limb syndrome in humans, called cardiac limb syndrome or Holt-Oram syndrome, is bilateral and involves the arms and hands. It is characterized by (p. 172)[2] thumbs that resemble fingers and hypoplasia of the forelimb bones; phocomelia (abnormally short and malformed limbs) may occur. This syndrome is known to be genetic, transmitted in an autosomal dominant fashion (only one parent need transmit the gene), and tends to be more severe in females (because affected male fetuses are more likely to be spontaneously aborted). We have not observed a comparable suite of characteristics in malformed frogs.

In the past 50 years, a handful of human infants are known to have been born with femoral hypoplasia-unusual faces syndrome (p. 170),[2] which presents as a suite of limb malformations that include "hypoplastic to absent femura and fibulae. Variable hypoplasia of humeri with restricted elbow movement, including synostosis. Club feet." The etiology is unknown; cases occur sporadically. The limbs of these infants can resemble the limbs of malformed frogs, but rarely is it the case with malformed frogs that all limbs are involved.

In humans, a rare limb condition known as radial aplasia (p. 176)[2] has affected > 20 patients. In this condition, caused by an autosomal recessive gene (i.e., both parents must pass on the gene), the radia (one of two forearm bones) are absent or hypoplasic (short). The condition is usually bilateral and associated with other defects in the arms and legs, hands and feet. Again, among malformed frogs, it is rarely the case that all limbs are involved.

Achrondrogenesis in humans is associated with severe micromelia (complete but small limbs), bony spurs, and a lack of ossification of vertebrae and other skeletal elements.[2] It is genetically based, transmitted in an autosomal recessive fashion. Achondrogenesis syndrome occurs sporadically and rarely, only a few dozen cases have been reported. A second type of achondrogenesis is characterized by marked limb reduction, especially distally, and polydactyly (six digits). There were 47 reported cases from among 5 Brazilian families.[2] Achrondrogenesis is an autosomal recessive genetic disorder. We do not see achondrogenesis among malformed frogs.

Dwarfism, or achondroplasia, occurs in humans with a frequency of about 1:10,000.[2] It is a genetically based, autosomal dominant malformation, with about 90% of cases representing new mutations. Older maternal age is a contributing factor. This syndrome is associated with a number of skeletal hypoplasias (shortenings) and delayed motor progress. Achondroplasia is unknown among malformed frogs, as are a variety of related chondroplasic, dysplasic, and dwarfism syndromes inherited through autosomal dominant or recessive genetic mechanisms.

One human congenital malformation that is reminiscent of frog malformations is called osteogenesis imperfecta tarda syndrome. It is characterized by bones with "thin cortices and sparse trabeculae with fragility leading to bowing of leg bones in severe cases" (p. 286).[2] The bent leg bones produced in this syndrome are reminiscent of the bowed long bones exhibited by the population of bullfrogs discovered by Pieter Johnson in California (see Figs. 2.32 and 3.15). There are differences. Humans with this autosomal dominant gene are short in stature, have limbs that are hyperextensible, skin that is thin and transparent, poor muscle development, and an increased tendency towards bleeding. Frogs do not show these features.

In humans, osteogenesis imperfecta congenita is an autosomal recessive malformation that produces "poorly mineralized long bones with multiple fractures and callus formation, especially in lower limbs" (p. 288).[2] The appearance of these bones is similar to some spongiform bony expansions seen in malformed frogs, the difference is that the expansions in frogs are typically limited to one leg, and on that leg at the site of limb truncation, while in humans, this skeletal condition affects limbs bilaterally, and all parts of all bones in each limb.

Non-genetic malformations of humans may be most relevant to the malformed frog problem since most frog malformations are likely to be caused by environmental (i.e., epigenetic) factors. In humans, epigenetic-produced malformations include fetal alcohol syndrome, which among many abnormalities includes joint, but not bone deficiencies. Rubella syndrome, caused by the Rubella virus, produces a suite of tissue defects including osseous lesions characterized by a course pattern of trabeculae and bone thinning. Drugs, such as the folic acid antagonist aminopterin (an abortion drug), the anticoagulant (blood thinner) coumadin, and the anticonvulsants (administered to control epileptic seizures) trimethadione and dilantin, affect soft tissues as well and skeletal tissues in systemic ways not usually seen in malformed frogs. In humans, these effects often result in fetal or neonatal death.

Polydactyly refers to too many fingers and toes; typically six digits are present. Incidence is roughly 1:500 live births.[2] In about 50% of cases, polydactyly is bilateral. A positive family history is present in approximately 30% of cases. Males are slightly more susceptible. Polydactyly can occur as an isolated case, or in association with other syndromes that include syndactyly, trisomy 21, trisomy 13, and hemimelia.

Syndactyly refers to a fusion of fingers or toes. In about 50% of cases, syndactyly is bilateral.[2] Syndactyly can occur as an isolated case, or it can be found in association with other syndromes such as polydactyly, cleft hands, constriction band (see below), and craniofacial syndromes such as Poland anomolad (see above) and Apert syndromes.

Congenital coxa vera includes all forms of decrease of the femoral neck angle present at birth.[2] This condition is assumed to be caused by either an embryonic limb bud abnormality or an abnormal intrauterine condition.

In all vertebrates, the nervous system—including the brain, spinal cord, and cranial and peripheral nerves—develops from neural ectoderm. The brain and spinal cord develop from a hollow tube that forms just above the bodies of the vertebrae (backbone). Failure of the neural tube to close in the pelvic area produces spina bifida, a common occurrence in humans (1:10), but rarely with functional effects (although paraplegia occurs in the most serious cases).[2] Failure of neural tube closure in the head region produces anencephaly, which is fatal. There is little evidence of neural tube defects in amphibians, and no amphibian malformations that we have observed were produced by neural tube defects. It is likely that neural tube defects do occur in amphibians but that they are fatal to embryos or newly hatched tadpoles, and therefore are not typically seen by field biologists.

By far, the most often encountered cause of congenital limb malformations in humans is "amniotic band" or "constriction band" syndrome,[27] which produces hemimely. Early in pregnancy, the uterus grows a web of filaments to support and secure the weight of the fetus. Occasionally one of the filaments pulls loose from the uterine wall, breaks, and then curls in reaction to the loss of tension. Sometimes it wraps around one of the fetal limbs, thus cutting off cellular migration and perhaps blood flow and preventing further development. With amniotic banding, no two cases are exactly alike. There are several features that are relatively consistent: syndactyly (webbed digits), distal ring constrictions, deformity of the nails, stunted growth of the small bones in the digits, limb length discrepancy, distal lymphedema (swelling), and congenital band indentations. Amniotic Band Syndrome (ABS) affects approximately 1:1,200 live births and is believed to be the cause of about 1:60 miscarriages. Up to 50% of ABS cases have other congenital anomalies including cleft lip and palate and clubfoot deformities. Hand and finger anomalies occur in up to 80% of cases. A characteristic of this type of accident is the presence of tiny "nubbins" of fingers or toes. Most amphibians, including all species that have demonstrated gross limb malformations, develop limbs outside of any maternal womb, and thus amniotic banding cannot be a cause of the missing limbs we observe.

It is possible, indeed likely, that factors affect developing frogs in one way and developing humans in another. The field of environmental contamination on human reproductive outcome is huge, often partisan, and largely beyond the scope of this book. Examples, based on the scientific literature, of the effects of pesticides on human reproduction can be found at www.pesticides.org.

CAUSES OF HUMAN MALFORMATIONS COMPARED WITH CAUSES OF AMPHIBIAN MALFORMATIONS

Again, the concern, expressed initially by all involved with the malformed frog phenomenon was that, because developmental processes tend to be conserved across vertebrates, what we could see happening to malformed frogs could also happen to humans.[28] A survey of human and frog malformations, however, shows little support for the literal interpretation of this position. Many human congenital malformations are due to chromosomal aberrations. With the exception of the paper by Lowcock and his colleagues,[29] very little is known about chromosomal abnormalities in amphibians (although chromosomal duplications appear to play a role in some amphibian speciation events[30]).[31] Further, many human limb malformations are due to an intrauterine phenomenon—constriction band syndrome—that cannot occur in free-living amphibian larvae. Finally, a list of known causes of human malformations includes several factors, such as parental age, seasonality, diet, diabetes, Rubella exposure, medical testing, in vitro fertilization, prescription and over-the-counter medications, illegal drug use, tobacco smoking, and alcohol consumption that at face value do not apply to amphibians (although amphibians can be exposed to much of what humans use through the habitation of sanitary sewers, storm sewers, retention ponds, and other downstream water systems).[32]

In contrast, exposure to the byproducts of the agricultural, leather, lumber, and health care industries, air pollution, water pollution including solvents and hazardous waste, pesticides, heavy metals, and endocrine disruptors known to affect humans will also affect amphibian development in adverse ways.[33] Again, the fact that exposures do not produce amphibian pathologies that resemble human pathologies should not be a reason to dismiss cause. After all, in the famous and well-worn analogy of canaries in the coal mine, concentrations of mine gases that kill canaries may instead cause people to have headaches and nausea.

SEVEN

SOLUTIONS

Come now, and let us reason together.

ISAIAH 53

Why is it that when human congenital malformations are addressed, the full suite of abnormalities and the full range of causes are considered, but when addressing malformations in frogs, the focus is typically only on single causes and their effects on hindlimbs, and then often only on hindlimb polymely? In almost every publication devoted to amphibian declines, we read that amphibians are "canaries in the coal mine," sensitive indicators of environmental health. To fully understand the role of amphibians as bioindicators, it would be wise to consider both the breadth and the depth of their sensitivity.

There is a curious dichotomy of answers in response to the question: "What causes frog malformations?" Much like the cartoon I show here (Fig. 7.1), lay people almost always respond with some variation of "It's them chemicals." Clearly, thalidomide, environmental disasters such as Love Canal, and no doubt a deeply rooted suspicion of pesticides, have permeated our collective consciousness. But ask the same question "What causes frog malformations?" to a scientist and the common response is "Parasites." The simple version of the parasite story has been told so well, so often, and in so many high-profile publications, that it seems that many scientists automatically respond as if the question of malformation causes has been settled once and for all.

So, what happens when a new population of malformed frogs is discovered? Is it immediately dismissed as unimportant because someone has

... I HOPE YOU'RE NOT GOING TO TRY TO BLAME THIS ONE ON ME...

FIGURE 7.1   Cartoon depicting the public perception that chemicals are causing amphibian malformations. From Simple Living. Used with permission of Simple Living and the cartoonist, Steve Babbert.

concluded that scientists have settled the issue and the cause is parasites—nothing to worry about, not much more than a little nutrient augmentation? Then, what happens two or three years later at the same place when epidemiologists discover a cluster of human congenital anomalies or disease? Such a congruence between American toad malformations and a cancer cluster in humans occurred several years ago in Knox County, Indiana, just south of where I currently live. Richard Feynman, speaking to the National Academy of Sciences, in 1955, said:[1]

> If we suppress all discussion, all criticism, proclaiming 'This is the answer, my friends; man is saved!' we will doom humanity for a long time to the . . . limits of our present imagination. It has been done so many times before.

The information presented in Chapter 5 demonstrates that every seriously considered cause of amphibian malformations has flaws as a universal theory.

Yes, it is true that nearly every identifiable *cause* has merit at some level—perhaps in the laboratory, or in the field locally, even regionally. It is also true that every cause fails as a general theory, due either to a lack of evidence or to contrary evidence. But, in fact, the *solution* to the malformed frog problem is clear enough, and to an ecologist it is as obvious as the white flash of a leopard frog's belly.

At the beginning of the phenomenon, Cindy Reinitz and her students had concerns that environmental health may be telling us something about human health. They were correct, and it would have been wise to have paid more attention to this link.

The most important and under-cited paper in all the modern malformed frog literature is the 1997 report by Les Lowcock and his colleagues,[2] which said, simply: "Malformed frogs have chromosomal damage."

Lowcock and his colleagues (p. 246) demonstrated that abnormal genetic profiles were more frequent in green frog populations exposed to pesticides. Lowcock's group showed that there was a significant difference in abnormal DNA profiles between animals exposed to agriculture and animals from control sites (p. 245), that coefficients of genetic variation were higher in malformed individuals (although sample sizes were low) than in normal individuals, and that a genotoxic effect was detectable in all field samples, with fields exposed to pesticides having the highest ratings (p. 246).

Not all malformed frogs have chromosomal damage. But even if chromosomal damage shows up in animals from only a handful of sites, it means that every new report of malformed frogs should raise deep concern. Which scenario is wisest: dismissing new reports as unimportant until a crisis arises (what happened in Indiana), or its opposite, treating each new report as a concern until it has been proven no risk exists? Human effects aside, from both ecological and conservation perspectives, all reports of malformations should be taken seriously, because no matter which cause is in play, the loss of individuals due to malformations represents a threat to host populations and their broader ecosystems.

There are two implications to the fact that malformed frogs have chromosomal damage; one is trivial, the other is critically important. The trivial implication is that where malformed frogs have chromosomal damage the cause is not likely to be parasites. Parasites are not known to disrupt chromosomal arrangements in either somatic or germinal cell lines. The second, gravely important implication is that if there is something out there causing chromosomal damage in frogs, it could be causing chromosomal damage in humans. Remember that a large number of severe congenital anomalies in humans are the result of chromosomal abnormalities (Chapter 6). In an interesting parallel, people exposed to the toxins at Love Canal had chromosomal damage.

It would be hard to overestimate the importance of Bob McKinnell's work. Bob is a scientific giant; he proved viruses cause cancer. And, while one of his friends happens to be the Emperor of Japan, Bob is not a fancy person; his study animals are northern leopard frogs from western Minnesota. Despite being in his 80s, he still drives out to the plains to study these leopard frog populations. Bob McKinnell was also one of three geneticists comprising an independent panel studying the evidence for human chromosomal damage at Love Canal, after the initial investigation got messy.[3] It went down like this: The Justice Department wanted to litigate. They collected evidence but did not provide sufficient control samples, claiming a lack of time and funding. The scientists did what they could with the material they had and wrote their report, considering it preliminary. The science was attacked ("False alarm caused by botched study"[4]) and the scientists mocked (with detractors claiming the results "exist only in the mind" of the lead scientist[5]). McKinnell's panel re-examined the evidence and concluded that the observed chromosomal aberrations were beyond normal limits, and that among the cytogenetic markers were unique large acentric fragments that could not be due to tissue sampling or processing artifacts. This opinion helped to re-establish balance.[6] Rumor has it that a lawsuit did the rest.

In a not-too-surprising parallel to the malformed frog investigation, while the world was bickering about the meaning of the chromosomal anomalies in people from Love Canal, the lead scientist of the Love Canal report noted (p. 754):[7]

> The cytogenetic study of the Love Canal population is totally overshadowed by the outcome of the last 18 pregnancies among the residents: two births were normal: nine children had birth defects: there were four spontaneous abortions and three stillbirths.

In a letter to Science defending the Love Canal study, Dr. Margery Shaw wrote (p. 752):[8]

Stochastic events that occur after exposure to mutagens, clastogens, carcinogens, and teratogens are not easy to explain. Chromosome damage is only one indicator in a series of poorly understood biological events that occur randomly in cells (and therefore in individuals) as a result of an external environmental insult. We cannot equate a ring chromosome in a lymphocyte with a cleft palate.

If the initial point of concern about amphibian malformations centered on what they might be telling us about threats to human health, if environmental health threats to humans are signaled by chromosomal abnormalities, and if at least some malformed frogs also exhibit chromosomal abnormalities, why hasn't the paper by Lowcock and his colleagues been cited more often, especially in high-profile review articles?

HOW TO CREATE A MALFORMED FROG

In Chapter 1, I present evidence for the view that most frog malformations are caused by epigenetic effects. As a reminder, epigenetic effects occur when genes and gene expression are normal, but some deviation from expected environmental circumstances occurs where and at the time when genes are being expressed. There are two basic ways to produce epigenetic effects: physical and chemical. Physical disruption of developing tissues can either eliminate or mechanically rearrange germinal cell lines. Trematode metacercarial cysts are known to do this. Failed predation, leech attacks, and other forms of physical irritation are also known to do this. Paradoxically, malformation types produced by mechanical disruption include both missing limbs (what you might expect) and extra limbs (what you might not expect). There can be considerable species specificity in response profiles (see Tables 5.1–5.3, 5.6, and 5.7) and wide geographic variation in the presence or intensity of the agent.

The second known way to produce epigenetic effects is to chemically alter either genetic expression or the products of genetic expression. In the laboratory, the direct application of chemicals such as retinoic acid can accomplish this, and while there has been much debate about retinoids in the published literature, it is important to realize that developmental biology is largely a chemical phenomenon. And, as Gardiner and Hoppe point out (p. 215):[9] ". . . the molecules involved in developmental signaling pathways are conserved structurally and functionally among all vertebrates . . .".

One way to imagine developmental pathways is as genetically determined chemical cascades, with chemicals appearing and disappearing at appropriate times and in appropriate quantities (developmental signaling pathways, in the parlance) to produce appropriate structures, as the genetic blueprint for development gets expressed. Viewed in this way, development can be affected by chemicals:

> required but missing;
>
> required but expressed at inappropriate times;
>
> present at the appropriate time but in inappropriate quantities;
>
> inappropriately present that block the actions of appropriate chemicals;
>
> inappropriately present that break apart or tear down formed or
> forming structures.

Chemicals acting in these ways will have direct effects on developing systems. As Gardiner and Hoppe note (p. 214):[9]

> There are many experimental studies demonstrating that chemical agents can disrupt normal embryonic development and induce congenital malformations. Collectively these studies form the basis for the scientific disciplines of developmental toxicology and teratology. In fact, much of what is known about normal development is derived from controlled, experimentally induced teratogenesis by both chemical and genetic means.

Chemicals can also have indirect effects, by affecting the immune system (if it has developed), the endocrine system (if it has developed), disrupting chromosomes, or through some combination of effects. Other environmental factors such as temperature, pH, and UV-B also affect the rate and nature of chemical effects. The role of stress in developing vertebrates, especially in amphibian embryos, which do not develop in utero (and therefore cannot be exposed to maternal endocrine and immune systems), remains largely unknown.[10]

Even with this uncertainty, by understanding some things about chemical effects we can address the following questions.

Q: *How can the same cause produce more than one malformed frog type?*

A: By affecting different developmental signaling pathways. Differences in timing and genetic variation within species can combine to produce diverse malformation types (consider the various thalidomide defects). Differences in timing and genetics can also produce different malformation types across species.

Q: *How can the same malformation type be produced by different causes?*

A: Different causes can affect the same developmental signaling pathways. This is true both within and across species. The more closely related any two species, the more recently they shared genomes (through common ancestry), and therefore the more recently they shared signaling pathways. Different causes can also affect the same developmental signaling pathways in species that inhabit various sites.

Q: *How is it that some sites produce unique malformation types?*

A: Frogs at different locations can have their developmental signaling pathways affected uniquely. The Trempealeau County, Wisconsin site; Switzerland County, Indiana site; and Pieter Johnson's California site (Chapter 3) are among the best examples of site-specific malformation types in the United States. Presumably either a unique factor, or a unique combination of factors, is present. These factors produce consistent malformations that are more similar among frogs sharing the wetland than they are to frogs from other sites.

EXCEEDING ECOLOGICAL TOLERANCES

In short, you can create a malformed frog by exceeding the ecological (including biochemical and physiological) tolerances of frog embryos or tadpoles. Exceed these tolerances greatly, especially in vital organs, and the animal dies. Exceed tolerances to a lesser degree and you get survival with malformations. The less critical a structure (limbs, for example, are not essential for tadpole viability), the more you can exceed tolerances and not directly affect survival.

Turn this concept upside down and you discover you can solve the malformed frog problem by bringing tolerances back in line.

THE SOLUTION TO THE MALFORMED
FROG PROBLEM

The strategy taken by most malformed frog researchers seems sound enough both in theory and in practice, and proceeds as follows: (1) develop an awareness; (2) identify causes; (3) effect solutions. As straightforward as this may seem, there have been two problems with this approach. First, the field has gotten so bogged down by squabbles over causes[11] that there have been few recommendations, not to mention fully formed plans, for solutions. As I mention in the Introduction: I know of no malformed frog hotspot that has been restored to ecological health through either the direct or indirect effects of biologists working on this issue.

Second, and paradoxically, considering where the effort has been spent, this is a field where you do not have to know cause in great detail in order to achieve a solution. That is, in following the logical flow from problem, to identifying cause, to effecting a solution, you do not need to spend much time considering cause. I will illustrate first with an example from human medicine. When a patient is brought to an emergency room with a headache, a stiff neck, purplish spots (petechiae in the parlance), and vomiting, the attending physician should not wait for test results to begin treatment. In the worst-case scenario the patient has bacterial meningitis, which can be fulminant and will likely be fatal in the time it takes to definitively establish cause. Instead, the attending physician should order an intravenous course of antibiotics, which should relieve the infection if it is bacterial meningitis, and do no harm (the physician's motto) if the origin is something else (e.g., viral or fungal meningitis, or a response to some other microbe), which will likely resolve on its own.

Similarly, there is little need to know details of the causes of amphibian malformations. This is a startling statement when we first hear it, but it is absolutely true, because the solution to most identified causes of malformation requires the same remediation. For example, let's say we believe that chemicals cause amphibian malformations. How does this happen? Typically, chemicals enter water systems through runoff and likely disrupt genetic expression through epigenetic mechanisms. That is, runoff contains chemicals that produce amphibian malformations. Fair enough. Now, let's say we believe parasites cause amphibian malformations. How does this happen? According to Pieter Johnson's body of work, nutrients entering water systems through runoff encourage plant growth, which supports increased snail populations, which support increased numbers of trematode parasites (*Ribeiroia ondatrae*), which in turn infect frogs, mechanically disrupt tissues, and produce malformations. That is, runoff contains nutrients that ultimately produce amphibian malformations. Again, fair enough. But note the common bottom line. No matter whether you believe that chemicals or parasites cause malformations: Control runoff and you control amphibian malformations.

This is a simple answer, I know, but I am not a fancy person, either. And while the solution is simple, the question "Can we can solve the problem?" is not. For example, if we could control runoff we could also eliminate "dead zones," areas of marine and freshwater habitats where the nutrient inputs from major river systems cause enormous algal blooms. The seasonal dead zone (March to October) at the mouth of the Mississippi River is huge, and often exceeds 13,000 km^2.[12] Eliminating these coastal dead zones will be difficult because of the scale of the problem (imagine trying to control point and non-point runoff throughout the Missouri, Mississippi, and Ohio river systems, our country's main agricultural areas). In comparison, controlling runoff into isolated wetland hotspots should be simple. Install non-agricultural vegetative buffers, reduce the overuse of both fertilizers and pesticides, fence off wetlands that need not be used by cattle, and install and maintain septic systems. After this is done, at each site conduct surveys to see how the amphibians respond. Such activities would be both personally and professionally rewarding, and could be the focus of local environmental groups.

In our 2003 paper (which includes Dan Sutherland, Dave Hoppe, and Pieter Johnson as co-authors), we anticipate portions of this discussion when we write (p. 260):[13]

> Recognizing that hotspots are altered wetlands, probably the quickest and least expensive way to reduce malformations is to recognize the nature of the alterations and take steps to eliminate them. For example, is part of the eutrophication and suspected chemical input to ROI due to leaky septic systems? This is easily tested and if so, these waste control systems should be upgraded. Is part of the eutrophication and suspected chemical input to CWB due to the utilization of this wetland by cattle? This is an especially instructive question, because while the debate on malformation causes has tended to focus on proximate causes (for example retinoids versus parasites) the fact is that both [parasites and retinoids] could be present, and both

could be caused by a single factor: cattle usage (nutrients produce eutrophi-cation which produces snails which produce trematodes; retinoids enter the water as a non-digested component of feed additives). In either case, para-sites or retinoids, the solution at CWB might be to remove the cattle.

In a later paper, Andy Blaustein and Pieter Johnson write (p. 93):[14]

If trematode parasitism is identified as the main cause of amphibian defor-mities, solutions should concentrate on . . . reducing nutrient inputs, more efficient use of fertilizers, and the reduction of cattle access to riparian areas.

In other words, you can reduce parasites by lowering the nutrient com-ponent of runoff.

Similarly, folks advocating a chemical cause for malformations might ad-vocate reducing the pesticide component of runoff. Flip these ideas over on their bellies and this leads to an interesting paradox. Staunch support-ers of a parasite cause for amphibian malformations must defend the po-sition that surface runoff does not contain chemical contamination, or that amphibians are somehow more immune, or not so vulnerable, to ex-posure to these chemicals. You get an inkling of this view when Blaustein and Johnson write (p. 64):[15] ". . . *Ribeiroia* is almost always found where deformed amphibians are present, whereas chemical pollutants are found much less frequently."

They later soften this position (p. 65):

Although parasitism by trematodes is the likeliest explanation for most outbreaks of amphibian deformities, it is certainly not the only cause. At times, water pollutants . . . may act alone [or] . . . set the stage for infection by weakening an amphibian's immune system . . . leaving the animal more vulnerable to a parasitic invasion.

In the same way, staunch defenders of a chemical cause for amphibian malformations are forced to claim that surface runoff does not contain nutrients, or that these nutrients are unimportant, do not cause eutrophi-cation, or do not produce ecological cascades that result in the amplified production of parasites. Both positions are not only tenuous, they are pre-posterous. Amphibian malformations are known from 71 species in the

United States alone (Appendix), with a geographic distribution that encompasses most of the country. To think that levels of nutrients and chemicals found in runoff do not: (1) vary geographically and seasonally, and (2) vary in relationship to each other geographically and seasonally, is to consider this issue from a minority perspective (see for example http:// toxics.usgs.gov/regional/emc.html). Further, among amphibians there is ample evidence for interspecific variation in responses to any given level of any particular chemical concentration, and to any given parasite density. In the face of this variation, advocating one particular scenario of general cause over any other will always lead to fact-based challenges; the kind that are impossible for scientists to counter.

THE NOT SO FUNNY LITTLE JOKE

While there is absolutely nothing funny about the malformed frog phenomenon, there has been an ongoing joke in the scientific investigation of malformed frogs. And with most good jokes, there is an element of deep truth revealed under the guise of humor. So, here's the malformed frog joke:

> While most of the published literature on the malformed frog issue has been devoted to causes of malformations, the solution to the malformed frog problem is largely independent of cause. All of the work that set out to solve the malformed frog problem by determining cause turns out to have not been so necessary after all.

Not really so funny, is it?

This conclusion—that the solution to the malformed frog problem is largely independent of cause—happens to be where the facts lead us. And rather than spend our time and grant money on yet another correlational analysis that will only serve to further polarize (if that is possible) the field, I suggest it will instead be better to go out to affected wetlands and start installing erosion control structures and measuring their effect.

TYING UP A LOOSE END

While the available data suggest that controlling runoff will reduce or eliminate most causes of amphibian malformations, we know of some places where malformations appear to be due to past land-use issues and on-site causes. For example, some essential utilities generate malformed frogs. Sewage treatment plants are often associated with malformed frogs (one young heavy equipment operator in east central Indiana quit her job transferring sludge because she didn't want what was happening to the frogs she was seeing to affect her and any children she might bear). Oil drilling retention ponds from the panhandle of Florida to northern Alberta also produce malformed amphibians. Further, aerial and atmospheric factors, such as chemical deposition and UV-B irradiance, are known to cause malformations. Both onsite and atmospheric causes of malformations will be difficult to remediate, and it will take some imagination to solve these problems.

CAUSES RECONSIDERED

Despite the fact that we can remediate malformed frog hotspots without knowing cause, I suspect that the search for causes will not go away; as a society, we love arguing theory, and we especially love dismissing contrary opinions. As the search for causes of amphibian malformations continues, any hypothesis is likely to be both correct and incorrect depending on scale: correct when addressing specific study sites; incorrect when extrapolating broadly. The reason for this is, as has always been known, that the malformed frog problem largely consists of local, site-specific issues. Hotspots are typically isolated wetlands, and no good reason has ever been given to support the assumption that causes that hold for distant sites will apply locally. In 2003, we wrote (p. 259):[13]

> Amphibian malformations have several causes and in the best of all possible worlds each hotspot would be carefully examined, the cause determined, and the source of the cause eliminated. This process is not only costly, but time consuming, and given the current funding crises experienced by governments at all levels, it is also not likely to happen anytime soon. Instead, it might be better to recognize that hotspots tend to be altered wetlands.

These alterations grade from what appears to be benign causes (wetlands simply being newly created and perhaps not having the buffering capacity of more mature ecosystems), through what is perhaps simple eutrophication (leading ultimately to parasites); through chemical inputs (leading to all sorts of problems including chromosomal abnormalities). While all categories of effect should be of concern, it is this latter category that should most affect human health.

SOCIETY'S ROLE

While science should, and must, lead the way in showing how to solve the malformed frog problem, the solutions themselves must come from a willing society. For example, at CWB, shouldn't the right of a home-owner to live without health concerns trump a farmer's need to water cattle? And shouldn't the farmer be given options so that he can continue to farm? Value systems, and in particular addressing value systems in cre-ative ways, then, becomes a central issue in solving the malformed frog problem. As a society, we have found these issues difficult to address.

In the Introduction, I note that many Americans seek moral guidance on environmental issues from Genesis 1:26. We should then ask what is moral about a society that permits the sorts of gross (in both senses of the word) malformations shown here? It can be argued that the application of Genesis 1:26 has produced, in the oft-sung words of Leonard Cohen, " . . . a cold and a broken Hallelujah."[16] Instead, when seeking Biblical perspectives on ecological stewardship, perhaps we should take a hard look in the mirror and adopt Jeremiah 2:7 ("And I brought ye into a plen-tiful country, to eat the fruit thereof and the goodness thereof; but when ye entered, ye defiled my land, and made mine heritage an abomina-tion"). Malformed frogs are unholy things. And while we imagine this is something we are doing unto others, let there be no doubt, as the chro-mosomes tell us, this is something we are doing unto ourselves.

SPECIES AFFECTED

These names . . . convey the sense of overwhelming
numbers.

MAYA LIN[1]

Amphibian malformations are not evenly distributed across taxa. As mentioned in Chapter 1, amphibian malformations tend to be frog malformations. A quick count shows that 52 of 105 U.S. frog species (50%) had
documented malformations compared with 19 of 188 U.S. salamander
species (10%).

The frog genera with the highest rates of documented malformations
include: *Ascaphus* (1 of 2 species or 50%), *Bufo* (10 of 23 species or 43%),
Acris (2 of 2 species or 100%), *Hyla* (7 of 10 species or 70%), *Osteopilus* (1
of 1 species or 100%), *Pseudacris* (8 of 14 species or 57%), *Gastrophryne* (2 of
2 species or 100%), *Scaphiopus* (2 of 3 species or 67%), *Spea* (1 of 4 species
or 25%), *Xenopus* (1 of 1 species or 100%), and *Rana* (17 of 31 species or
55%). The salamander genera with the highest rates of documented malformations include *Ambystoma* (8 of 17 species or 47%, plus unisexual hybrids), *Notophthalmus* (1 of 3 species or 33%), and *Taricha* (2 of 3 species
or 67%).

The frog genera *Dendrobates, Pternohyla, Smilisca, Eleutherodactylus, Leptodactylus, Hypopachus,* and *Rhynophrynus* have no documented malformations. Likewise, the salamander genera *Amphiuma, Cryptobranchus,
Dicamptodon, Aneides, Batrachoseps, Ensatina, Gyrinophilus, Haideotriton,
Hemidactylium, Hydromantes, Phaeognathus, Pseudotriton, Stereochilus, Typhlotriton, Necturus, Rhyacotriton, Pseudobranchus,* and *Siren* have no documented malformations. The salamander genera with few documented

malformations include *Desmognathus* (1 of 19 species or 5%), *Eurycea* (2 of 27 species or 7%), and *Plethodon* (5 of 54 species or 9%).

The following is a listing of North American amphibian species;[2] names of species with documented malformations are in boldface type. The number of animals that I have radiographed appears to the right of each species name. In addition to these animals, 74 unknown *Bufo*, 6 unknown *Acris*, 74 unknown *Rana*, 38 unknown frogs, 1 unknown *Ambystoma*, and 4 unknown salamanders were radiographed, for a total of 2,377 images.

FROGS: ORDER ANURA

FAMILY ASCAPHIDAE

Ascaphus montanus, Montana (Mountain) Tailed Frog

Ascaphus truei*, (Coastal) Tailed Frog[3]   1

FAMILY BUFONIDAE

Bufo alvarius, Colorado River Toad

Bufo americanus, American Toad[2,4,5,6]   70

Bufo baxteri, Wyoming Toad

Bufo boreas, Western Toad[2,7]   34

Bufo californicus, Arroyo Toad[2]

Bufo canorus, Yosemite Toad

Bufo cognatus, Great Plains Toad

Bufo debilis, Green Toad

Bufo exsul, Black Toad

Bufo fowleri, Fowler's Toad[2,8]   7

Bufo hemiophrys, Canadian Toad[2]

Bufo houstonensis, Houston Toad

Bufo marinus, Marine Toad or Cane Toad[2,9]   96

Bufo microscaphus, Arizona Toad

Bufo nebulifer, Coastal-Plain Toad[2,10]

Bufo nelsoni, Amargosa Toad

Bufo punctatus, Red-Spotted Toad[2]

Bufo quercicus, Oak Toad

Bufo retiformis, Sonoran Green Toad

Bufo speciosus, Texas Toad

Bufo terrestris, Southern Toad[2]

Bufo velatus, East Texas Toad

Bufo woodhousii, Woodhouse's Toad[2]   I

FAMILY DENDROBATIDAE

Dendrobates auratus, Green and Black Dart-Poison Frog

FAMILY HYLIDAE

Acris crepitans, Northern Cricket Frog[2,11]   53

Acris gryllus, Southern Cricket Frog[2]   13

Hyla andersonii, Pine Barrens Treefrog

Hyla arenicolor, Canyon Treefrog[2]

Hyla avivoca, Bird-Voiced Treefrog

Hyla chrysoscelis, Cope's Gray Treefrog[2]   11

Hyla cinerea, Green Treefrog[2]   3

Hyla femoralis, Pine Woods Treefrog[12]   3

Hyla gratiosa, Barking Treefrog

Hyla squirella, Squirrel Treefrog[12]   4

Hyla versicolor, Eastern Gray Treefrog[2]   I

Hyla wrightorum (eximia), Arizona Treefrog[2]

Osteopilus septentrionalis, Cuban Treefrog[2]

Pseudacris brachyphona, Mountain Chorus Frog

Pseudacris brimleyi, Brimley's Chorus Frog

Pseudacris cadaverina, California Treefrog[2]

Pseudacris clarkii, Spotted Chorus Frog

Pseudacris crucifer, Spring Peeper[2]   6

Pseudacris feriarum, Southeastern Chorus Frog[12]   17

Pseudacris maculata, Boreal Chorus Frog[2]

Pseudacris nigrita, Southern Chorus Frog

Pseudacris ocularis, Little Grass Frog[12]   2

Pseudacris ornata, Ornate Chorus Frog[2]

***Pseudacris regilla*, Pacific Treefrog**[2,3,13]   65

Pseudacris streckeri, Strecker's Chorus Frog

***Pseudacris triseriata*, Western Chorus Frog**[2]   I

Pternohyla fodiens, Lowland Burrowing Treefrog

Smilisca baudinii, Mexican Treefrog

FAMILY LEPTODACTYLIDAE

Eleutherodactylus augusti, Barking Frog

Eleutherodactylus coqui, Coqui

Eleutherodactylus cystignathoides, Rio Grande Chirping Frog

Eleutherodactylus guttilatus, Spotted Chirping Frog

Eleutherodactylus marnockii, Cliff Chirping Frog

Eleutherodactylus martinicensis, Martinique Greenhouse Frog

Eleutherodactylus planirostris, Greenhouse Frog

Leptodactylus fragilis, White-Lipped Frog

FAMILY MICROHYLIDAE

***Gastrophryne carolinensis*, Eastern Narrow-Mouthed Toad**[2]   4

***Gastrophryne olivacea*, Western Narrow-Mouthed Toad**[2]

Hypopachus variolosus, Sheep Frog

FAMILY PELOBATIDAE

***Scaphiopus couchii*, Couch's Spadefoot**[2]

***Scaphiopus holbrookii*, Eastern Spadefoot**[2]

Scaphiopus hurterii, Hurter's Spadefoot

***Spea bombifrons*, Plains Spadefoot**[3]

Spea hammondii, Western Spadefoot

Spea intermontana, Great Basin Spadefoot

Spea multiplicata, Mexican Spadefoot

FAMILY PIPIDAE

***Xenopus laevis*, African Clawed Frog**[2,14]

Rana areolata, Crawfish Frog

Rana aurora, Northern Red-Legged Frog[2]   66

Rana berlandieri, Rio Grande Leopard Frog

Rana blairi, Plains Leopard Frog[2]   29

Rana boylii, Foothill Yellow-Legged Frog[2,15]

Rana capito, Gopher Frog

Rana cascadae, Cascade Frog[2]   11

Rana catesbeiana, American Bullfrog[2,3,4,16]   45

Rana chiricahuensis, Chiricahua Leopard Frog

Rana clamitans, Green Frog[2,17]   97

Rana draytonii, California Red-Legged Frog[18]

Rana fisheri, Vegas Valley Leopard Frog

Rana grylio, Pig Frog[2]

Rana heckscheri, River Frog

Rana luteiventris, Columbia Spotted Frog[2]   3

Rana muscosa, Mountain Yellow-Legged Frog[2]

Rana okaloosae, Florida Bog Frog

Rana onca, Relict Leopard Frog

Rana palustris, Pickerel Frog[2,19]   5

Rana pipiens, Northern Leopard Frog[2,20]   709

Rana pretiosa, Oregon Spotted Frog[2]   2

Rana rugosa, Wrinkled Frog

Rana sierrae, Sierra Nevada Yellow-Legged Frog

Rana septentrionalis, Mink Frog[2,21]   6

Rana sevosa, Dusky Gopher Frog

Rana sphenocephala, Southern Leopard Frog[2]   402

Rana subaquavocalis, Ramsey Canyon Leopard Frog

Rana sylvatica, Wood Frog[2,4]   397

Rana tarahumarae, Tarahumara Frog

Rana virgatipes, Carpenter Frog[12]   2

Rana yavapaiensis, Lowland Leopard Frog

FAMILY RHINOPHRYNIDAE

Rhinophrynus dorsalis, Burrowing Toad

SALAMANDERS: ORDER CAUDATA
FAMILY AMBYSTOMIDAE

Ambystoma annulatum, Ringed Salamander

Ambystoma barbouri, Streamside Salamander

Ambystoma bishopi, Frosted Flatwoods Salamander

Ambystoma californiense, California Tiger Salamander

Ambystoma cingulatum, Reticulated Flatwoods Salamander

Ambystoma gracile, Northwestern Salamander

Ambystoma jeffersonianum, Jefferson Salamander[2]

Ambystoma laterale, Blue-Spotted Salamander[2,22]   2

Ambystoma mabeei, Mabee's Salamander

Ambystoma macrodactylum, Long-Toed Salamander[2,23]

Ambystoma maculatum, Spotted Salamander[2,24]

Ambystoma opacum, Marbled Salamander[2]

Ambystoma platineum, Silvery Salamander

Ambystoma talpoideum, Mole Salamander[25]

Ambystoma texanum, Small-Mouthed Salamander[2]

Ambystoma tigrinum, Tiger Salamander[2,26]   12

Ambystoma tremblayi, Tremblay's Salamander

Ambystoma sp., unisexual hybrids in the *Ambystoma jeffersonianum* complex

FAMILY AMPHIUMIDAE

Amphiuma means, Two-Toed Amphiuma

Amphiuma pholeter, One-Toed Amphiuma

Amphiuma tridactylum, Three-Toed Amphiuma

FAMILY CRYPTOBRANCHIDAE

Cryptobranchus alleganiensis, Hellbender

FAMILY DICAMPTODONTIDAE

Dicamptodon aterrimus, Idaho Giant Salamander

Dicamptodon copei, Cope's Giant Salamander

Dicamptodon ensatus, California Giant Salamander

Dicamptodon tenebrosus, Coastal Giant Salamander

FAMILY PLETHODONTIDAE

Aneides aeneus, Green Salamander

Aneides ferreus, Clouded Salamander

Aneides flavipunctatus, Black Salamander

Aneides hardii, Sacramento Mountains Salamander

Aneides lugubris, Arboreal Salamander

Aneides vagrans, Wandering Salamander

Batrachoseps aridus, Desert Slender Salamander

Batrachoseps attenuatus, California Slender Salamander

Batrachoseps campi, Inyo Mountains Salamander

Batrachoseps diabolicus, Hell Hollow Slender Salamander

Batrachoseps gabrieli, San Gabriel Mountains Slender Salamander

Batrachoseps gavilanensis, Gabilan Mountains Slender Salamander

Batrachoseps gregarius, Gregarious Slender Salamander

Batrachoseps incognitus, San Simeon Slender Salamander

Batrachoseps kawia, Sequoia Slender Salamander

Batrachoseps luciae, Santa Lucia Mountains Slender Salamander

Batrachoseps major, Garden Slender Salamander

Batrachoseps minor, Lesser Slender Salamander

Batrachoseps nigriventris, Black-Bellied Slender Salamander

Batrachoseps pacificus, Channel Islands Slender Salamander

Batrachoseps regius, Kings River Slender Salamander

Batrachoseps relictus, Relictual Slender Salamander

Batrachoseps robustus, Kern Plateau Salamander

Batrachoseps simatus, Kern Canyon Slender Salamander

Batrachoseps stebbinsi, Tehachapi Slender Salamander

Batrachoseps wrighti, Oregon Slender Salamander

Desmognathus abditus, Cumberland Dusky Salamander

Desmognathus aeneus, Seepage Salamander

Desmognathus apalachicolae, Apalachicola Dusky Salamander

Desmognathus auriculatus, Southern Dusky Salamander

Desmognathus brimleyorum, Ouachita Dusky Salamander

Desmognathus carolinensis, Carolina Mountain Dusky Salamander

Desmognathus conanti, Spotted Dusky Salamander

Desmognathus folkertsi, Dwarf Black-Bellied Salamander

Desmognathus fuscus, Northern Dusky Salamander[2,27]

Desmognathus imitator, Imitator Salamander

Desmognathus marmoratus, Shovel-Nosed Salamander

Desmognathus monticola, Seal Salamander

Desmognathus ochrophaeus, Allegheny Mountain Dusky Salamander

Desmognathus ocoee, Ocoee Salamander

Desmognathus orestes, Blue Ridge Dusky Salamander

Desmognathus quadramaculatus, Black-Bellied Salamander

Desmognathus santeetlah, Santeetlah Dusky Salamander

Desmognathus welteri, Black Mountain Salamander

Desmognathus wrighti, Pigmy Salamander

Ensatina eschscholtzii, Ensatina

Eurycea aquatica, Dark-Sided Salamander

Eurycea bislineata, Northern Two-Lined Salamander[2]

Eurycea sp. 3, Chamberlain's Dwarf Salamander

Eurycea chisholmensis, Salado Salamander

Eurycea cirrigera, Southern Two-Lined Salamander

Eurycea guttolineata, Three-Lined Salamander

Eurycea junaluska, Junaluska Salamander

Eurycea latitans, Cascade Caverns Salamander

Eurycea longicauda, Long-Tailed Salamander

Eurycea lucifuga, Cave Salamander[2]

Eurycea multiplicata, Many-Ribbed Salamander

Eurycea nana, San Marcos Salamander

Eurycea naufragia, Georgetown Salamander

Eurycea neotenes, Texas Salamander

Eurycea pterophila, Fern Bank Salamander

Eurycea quadridigitata, Dwarf Salamander

Eurycea rathbuni, Texas Blind Salamander

Eurycea robusta, Blanco Blind Salamander

Eurycea sosorum, Barton Springs Salamander

Eurycea tonkawae, Jollyville Plateau Salamander

Eurycea tridentifera, Comal Blind Salamander

Eurycea troglodytes, Valdina Farms Salamander

Eurycea tynerensis, Oklahoma Salamander

Eurycea waterlooensis, Austin Blind Salamander

Eurycea wilderae, Blue Ridge Two-Lined Salamander

Eurycea sp. 1, Comal Springs Salamander

Eurycea sp. 2, Pedernales Springs Salamander

Gyrinophilus gulolineatus, Berry Cave Salamander

Gyrinophilus palleucus, Tennessee Cave Salamander

Gyrinophilus porphyriticus, Spring Salamander

Gyrinophilus subterraneus, West Virginia Spring Salamander

Haideotriton wallacei, Georgia Blind Salamander

Hemidactylium scutatum, Four-Toed Salamander

Hydromantes brunus, Limestone Salamander

Hydromantes platycephalus, Mt. Lyell Salamander

Hydromantes shastae, Shasta Salamander

Phaeognathus hubrichti, Red Hills Salamander

Plethodon ainsworthi, Bay Springs Salamander

Plethodon albagula, Western Slimy Salamander[28]

Plethodon amplus, Blue Ridge Gray-Cheeked Salamander

Plethodon angusticlavius, Ozark Zigzag Salamander

Plethodon asupak, Scott Bar Salamander

Plethodon aureolus, Tellico Salamander

Plethodon caddoensis, Caddo Mountain Salamander

Plethodon chattahoochee, Chattahoochee Slimy Salamander

Plethodon cheoah, Cheoah Bald Salamander

Plethodon chlorobryonis, Atlantic Coast Slimy Salamander

Plethodon cinereus, Eastern Red–Backed Salamander[2,29]

Plethodon cylindraceus, White-Spotted Slimy Salamander

Plethodon dorsalis, Northern Zigzag Salamander

Plethodon dunni, Dunn's Salamander

Plethodon electromorphus, Northern Ravine Salamander

Plethodon elongatus, Del Norte Salamander

Plethodon fourchensis, Fourche Mountain Salamander

Plethodon glutinosus, Northern Slimy Salamander[2,30]

Plethodon grobmani, Southeastern Slimy Salamander

Plethodon hoffmani, Valley and Ridge Salamander

Plethodon hubrichti, Peaks of Otter Salamander

Plethodon idahoensis, Coeur d'Alene Salamander

Plethodon jordani, Jordan's Salamander or Red-Cheeked Salamander

Plethodon kentucki, Cumberland Plateau Salamander[31]

Plethodon kiamichi, Kiamichi Slimy Salamander

Plethodon kisatchie, Louisiana Slimy Salamander

Plethodon larselli, Larch Mountain Salamander

Plethodon meridianus, South Mountain Gray-Cheeked Salamander

Plethodon metcalfi, Southern Gray-Cheeked Salamander

Plethodon mississippi, Mississippi Slimy Salamander

Plethodon montanus, Northern Gray-Cheeked Salamander

Plethodon neomexicanus, Jemez Mountains Salamander[32]

Plethodon nettingi, Cheat Mountain Salamander

Plethodon ocmulgee, Ocmulgee Slimy Salamander

Plethodon ouachitae, Rich Mountain Salamander

Plethodon petraeus, Pigeon Mountain Salamander

Plethodon punctatus, Cow Knob Salamander

Plethodon richmondi, Southern Ravine Salamander

Plethodon savannah, Savannah Slimy Salamander

Plethodon sequoyah, Sequoyah Slimy Salamander

Plethodon serratus, Southern Red-Backed Salamander

Plethodon shenandoah, Shenandoah Salamander

Plethodon shermani, Red-Legged Salamander

Plethodon stormi, Siskiyou Mountains Salamander

Plethodon teyahalee, Southern Appalachian Salamander

Plethodon vandykei, Van Dyke's Salamander

Plethodon variolatus, South Carolina Slimy Salamander

Plethodon vehiculum, Western Red-Backed Salamander

Plethodon ventralis, Southern Zigzag Salamander

Plethodon virginia, Shenandoah Mountain Salamander

Plethodon websteri, Webster's Salamander

Plethodon wehrlei, Wehrle's Salamander

Plethodon welleri, Weller's Salamander

Plethodon yonahlossee, Yonahlossee Salamander

Pseudotriton montanus, Mud Salamander

Pseudotriton ruber, Red Salamander

Stereochilus marginatus, Many-Lined Salamander

Typhlotriton spelaeus, Grotto Salamander

FAMILY PROTEIDAE

Necturus alabamensis, Black Warrior Waterdog

Necturus beyeri, Gulf Coast Waterdog

Necturus lewisi, Neuse River Waterdog

Necturus maculosus, Mudpuppy

Necturus punctatus, Dwarf Waterdog

Necturus cf. beyeri, Loding's Waterdog

FAMILY RHYACOTRITONIDAE

Rhyacotriton cascadae, Cascade Torrent Salamander

Rhyacotriton kezeri, Columbia Torrent Salamander

Rhyacotriton olympicus, Olympic Torrent Salamander

Rhyacotriton variegatus, Southern Torrent Salamander

Notophthalmus meridionalis, Black-Spotted Newt

Notophthalmus perstriatus, Striped Newt

Notophthalmus viridescens, Eastern Newt[2]

Taricha granulosa, Rough-Skinned Newt[2,33]

Taricha rivularis, Red-Bellied Newt

Taricha torosa, California Newt[2]

FAMILY SIRENIDAE

Pseudobranchus axanthus, Southern Dwarf Siren

Pseudobranchus striatus, Northern Dwarf Siren

Siren intermedia, Lesser Siren

Siren lacertina, Greater Siren

Siren texana, Rio Grande Siren(s)

These taxonomic patterns reveal several ecological and methodological trends, including some that remain open to interpretation. The first trend, which will come as no surprise to anyone who has followed this issue, is that aquatic species are most at risk. The genera with the highest rates of documented malformations include *Ascaphus, Bufo, Acris, Hyla, Osteopilus, Pseudacris, Gastrophryne, Scaphiopus, Spea, Xenopus, Rana, Ambystoma, Notophthalmus,* and *Taricha.* All species in these genera have aquatic larvae. The genera with the lowest rates of documented malformations include *Eleutherodactylus, Leptodactylus, Aneides, Batrachoseps, Ensatina, Hydromantes, Phaeognathus,* and *Plethodon.* All species in these genera are terrestrial. Dave Hoppe has found this same pattern of risk associated with aquatic habitats within the genus *Rana,* with the most aquatic species having the highest rates and most severe malformations[34] (but see Johnson and colleagues [p. 348]).[35] From the start of the malformed frog phenomenon in the mid-1990s, malformed frog reports carried with them the assumption that "something was in the water."[36]

Not all aquatic genera are at risk. The genera *Amphiuma, Cryptobranchus, Dicamptodon, Haideotriton, Necturus, Rhyacotriton, Pseudobranchus,* and *Siren* have no reported malformations. One reason may be that these animals inhabit larger water bodies—rivers, streams, and lakes—than species in more affected genera. Large bodies of water provide more opportunity for dilution (the euphonious solution to pollution) and less opportunity for aquatic plant growth than do the restricted, shallow waters of wetlands. A second reason why malformations in these animals are rarely observed, which also must be true, is that these animals are rarely encountered. This is because: (1) as noted above they inhabit large water bodies and (2) they do not metamorphose into terrestrial adults that crawl onto land where we can easily see them.

The genera *Desmognathus* and *Eurycea* tend to be streamside salamanders. They are common and speciose and likely show up here as having low rates of malformations simply because they have low rates of malformations.

Other genera, including the frogs *Dendrobates*, *Pternohyla*, *Smilisca*, *Hypopachus*, and *Rhynophrynus* and the salamanders *Gyrinophilus*, *Hemidactylium*, *Pseudotriton*, *Stereochilus*, *Typhlotriton* (*Eurycea*), have no documented malformations. These animals also tend to be rarely seen for one or more reasons including: (1) they tend to live underground; (2) they are otherwise secretive; (3) they are rare; and (4) they are restricted taxonomically or geographically. It is difficult to overestimate the secretive nature of some species of amphibians. For example, to this day, scientists have not observed egg masses in some species of *Batrachoseps*—we have no first-hand evidence of clutch size in these animals.[2] It is no exaggeration to point out that we humans likely see more northern leopard frogs every day than we have ever seen individuals of some species of *Batrachoseps* in the entire history of mankind.

NOTES

PREFACE

1. Lin, M. 2000. Boundaries. Simon and Schuster, New York.

2. See for example, Lannoo, M.J. (Ed.) 2005. Amphibian Declines: The Conservation Status of United States Species. University of California Press, Berkeley, California.

3. Lannoo, M.J. 2000. Conclusions drawn from the malformity and disease session, Midwest Declining Amphibians Conference, 1998. Pp. 212–216. *In* Kaiser, H., G.S. Casper and N.P. Bernstein (Eds.), Investigating Amphibian Declines: Proceedings of the 1998 Declining Amphibians Conference. Journal of the Iowa Academy of Science, Volume 107, Cedar Falls, Iowa.

4. Souder, W. 2000. A Plague of Frogs: The Horrifying True Story. Hyperion Press, New York.

5. Lannoo, M.J., D.R. Sutherland, P. Jones, D. Rosenberry, R.W. Klaver, D.M. Hoppe, P.T.J. Johnson, K.B. Lunde, C. Facemire and J.M. Kapfer. 2003. Multiple causes for the malformed frog phenomenon. pp. 233–262 *in* Linder, G., S. Krest, D. Sparling and E. Little (Eds.), Multiple Stressor Effects in Relation to Declining Amphibian Populations. American Society for Testing Materials International, West Conshohocken, Pennsylvania.

6. Mecchi, I., J. Roberts and L. Woolverton. 1994. The Lion King. Walt Disney Studios, Anaheim, California.

ACKNOWLEDGMENTS

1. Steinbeck, J. and E.F. Ricketts. 1941. Sea of Cortez: A Leisurely Journal of Travel and Research. Viking Press, New York.

INTRODUCTION

1. Gould, S.J. 1983. An Urchin in the Storm. W.W. Norton and Company, New York, p. 218.

2. Souder, W. 2000. A Plague of Frogs: The Horrifying True Story. Hyperion Press, New York.

3. Thalidomide was first marketed in West Germany in 1956 as a hypnotic, as well as a treatment for morning sickness. The drug was successfully launched in a number of countries, including Britain in 1958, and was generally accepted as safe and effective. However, in 1961 it became clear that if thalidomide was taken during pregnancy it could cause phocomelia, a rare congenital abnormality in which the long bones fail to develop. Volpe and Rosenbaum write that more than 6,000 thalidomide babies were born in West Germany and at least 1,000 more in other countries (p. 4; reference immediately below). The United States was largely spared this disaster because of the courageous decision of Frances O. Kelsey, of the U.S. Food and Drug Administration, to not register thalidomide despite intense pressure. The difference one brave person can make is an optimistic and repeating theme played over and over again throughout the history of mankind.

   Volpe, E.P. and P.A. Rosenbaum. 2000. Understanding Evolution. Sixth Edition. McGraw-Hill, New York.

4. Love Canal "And then there were the birth defects." As I discuss in Chapter 8, at one point the author of the primary study wrote (Science 209:754): ". . . the outcome of the last 18 pregnancies among the residents [has been] two births were normal: nine children had birth defects: there were four spontaneous abortions and three stillbirths."

5. See http://armi.usgs.gov/

6. von Drehle, D. 2003. Triangle: The Fire That Changed America. Atlantic Monthly Press, New York.

7. Backes, D. 1997. The Life of Sigurd F. Olson. University of Minnesota Press, Minneapolis, Minnesota. This quote is from Leopold's lecture "The Critical Approach to Wildlife" (see Meine, C. 1988. Aldo Leopold: His Life and Work. University of Wisconsin Press, Madison, Wisconsin.)

8. Barry, J.M. 2004. The Great Influenza: The Epic Story of the Deadliest Plague in History. Penguin, London, England. Part 1: The Warriors: Chapter 1.

9. At the Indiana University School of Medicine, where I teach, we have adopted a competency-based curriculum, that centers on the following: effective communication; basic clinical skills; using science to guide diagnosis, management, therapeutics, and prevention; lifelong learning; self-awareness, self-care, and personal growth; social and community contexts of health care; moral reasoning and ethical judgment; problem solving; and professionalism and role recognition.

10. Maclean, N. 1994. Young Men and Fire. University of Chicago Press, Chicago, Illinois.

11. Gould, S.J. 1998. Leonardo's Mountain of Clams and the Diet of Worms. Three Rivers Press, New York.

12. Gould, S.J. 1983. An Urchin in the Storm. W.W. Norton and Company, New York.

13. Marshall, E. 2006. Royal Society takes a shot at ExxonMobil. Science 313:1871.

14. Frost, R. 1913. Mowing. *From* A Boy's Will. David Nutt, London, England.

15. Popper, K. 1935. Logik der Forschung. Springer-Verlag, Vienna, Austria. First English edition published as The Logic of Scientific Discovery in 1959 by Hutchison and Company, and reprinted in paperback by Routledge, New York. Copyright 2002, The Estate of Karl Popper.

16. Gould, S.J. 2000. Lying Stones of Marrakech. Harmony Books, New York.

17. Gonzales, L. 2004. Deep Survival: Who Lives, Who Dies, and Why. W.W. Norton and Company, New York.

18. Rodger, K.A. 2006. Breaking Through: Essays, Journals, and Travelogues of Edward F. Ricketts. University of California Press, Berkeley, California.

CHAPTER 1: WHAT IS AN AMPHIBIAN MALFORMATION?

1. Pollan, M. 2001. Botany of Desire. Random House, New York.

2. Much of the information I present here on vertebrate development is well known and well represented in texts on embryology. In assembling this chapter I have relied on a variety of sources that were important to my initial learning of this material, including:

Le Douarin, N. 1984. The Neural Crest. Cambridge University Press, Cambridge, England.

Moore, K.L. 1988. The Developing Human. Saunders, Philadelphia, Pennsylvania.

Noden, D.M. and A. de Lahunta. 1985. The Embryology of Domestic Animals: Developmental Mechanisms and Malformations. Williams and Wilkins, Baltimore, Maryland.

Oppenheimer, S.B. and G. Lefevre, Jr. 1984. Introduction to Embryonic Development. Second Edition. Allyn and Bacon, Inc., Boston, Massachusetts.

3. In July 2003, two cinematographers, Alastair MacEwen and Ralph Bower, on assignment from The National Geographic Society collected this large (95.5 mm SUL) male green and blue (axanthic) northern leopard frog along the shore of Sunken Lake in Dickinson County, Iowa. Axanthism is a genetic mutation producing skin that is literally without yellow, and therefore blue, and is described by H.B. Bechtel (1995, Reptile and Amphibian Variants: Colors, Patterns, and Scales. Krieger Publishing Company, Malabar, Florida) as a hereditary defect of xanthophore pigment metabolism in which carotenoids and pteridines are substantially reduced. Among amphibians, axanthism is widely distributed and exhibited at the highest frequencies in North America in the Upper Midwest, New England, and southeastern Canada. Axanthic frogs occur at a rate of about 1:30,000 animals, the rate of expression of albinos.

4. Hoppe, D.M. and R.G. McKinnell. 1991. Minnesota's mutant leopard frogs. Minnesota Volunteer 1991:56–63.

McKinnell, R.G. and D.M. Hoppe. 2005. Monitoring pigment pattern morphs of northern leopard frogs. pp. 328–337 in Lannoo, M.J. (Ed.), Amphibian Declines: The Conservation Status of United States Species. University of California Press, Berkeley, California.

5. Johnson, P.T.J., K.B. Lunde, E.G. Ritchie and A.E. Launer. 1999. The effect of trematode infection on amphibian limb development and survivorship. Science 284:802–804.

Johnson, P.T.J., K.B. Lunde, R.W. Haight, J. Bowerman and A.R. Blaustein. 2001. Ribeiroia ondatrae (Trematoda: Digena) infection induces severe limb malformations in western toads (Bufo boreas). Canadian Journal of Zoology 79:370–379.

Johnson, P.T.J., K.B. Lunde, E.G. Ritchie, J.K. Reaser and A.E. Launer. 2001. Morphological abnormality patterns in a California amphibian community. Herpetologica 57:336–352.

Johnson, P.T.J., K.B. Lunde, E.M. Thurman, E.G. Ritchie, S.N. Wray, D.R. Sutherland, J.M. Kapfer, T.J. Frest, J. Bowerman and A.R. Blaustein. 2002. Parasite (Ribeiroia ondatrae) infection linked to amphibian malformations in the western United States. Ecological Monographs 72:151–168.

Johnson, P.T.J. and K.B. Lunde. 2005. Parasite infection and limb malformations: A growing problem in amphibian conservation. pp. 124–138 in Lannoo,

M.J. (Ed.), Amphibian Declines: The Conservation Status of United States Species. University of California Press, Berkeley, California.

    Johnson, P.T.J., J.M. Chase, K. L. Dosch, R.B. Hartson, J.A. Gross, D.J. Larson, D.R. Sutherland and S.R. Carpenter. 2007. Aquatic eutrophication promotes pathogenic infection in amphibians. Proceedings of the National Academy of Science 104:15781–15786.

6. Johnson, P.T.J., E.R. Preu, D.R. Sutherland, J.M. Romansic, B. Han and A.R. Blaustein. 2006. Adding infection to injury: Synergistic effects of predation and parasitism on amphibian malformations. Ecology 87:2227–2235.

7. Gardiner, D.M. and D.M. Hoppe. 1999. Environmentally induced limb malformations in mink frogs (*Rana septentrionalis*). Journal of Experimental Zoology 284:207–216.

    Lannoo, M.J. 2000. Conclusions drawn from the malformity and disease session, Midwest Declining Amphibians Conference, 1998. pp. 212–216 *in* Kaiser, H., G.S. Casper and N.P. Bernstein (Eds.), Investigating Amphibian Declines: Proceedings of the 1998 Declining Amphibians Conference. Journal of the Iowa Academy of Science, Volume 107, Cedar Falls, Iowa.

    Lannoo, M.J., D.R. Sutherland, P. Jones, D. Rosenberry, R.W. Klaver, D.M. Hoppe, P.T.J. Johnson, K.B. Lunde, C. Facemire and J.M. Kapfer. 2003. Multiple causes for the malformed frog phenomenon. pp. 233–262 *in* Linder, G., S. Krest, D. Sparling and E. Little (Eds.), Multiple Stressor Effects in Relation to Declining Amphibian Populations. American Society for Testing Materials International, West Conshoshocken, Pennsylvania.

    Hoppe, D.M. 2000. History of Minnesota frog abnormalities: do recent findings represent a new phenomenon? pp. 86–89 *in* Kaiser, H., G.S. Casper and N. Bernstein (Eds.), Investigating Amphibian Declines: Proceedings of the 1998 Midwest Declining Amphibians Conference. Iowa Academy of Science, Volume 107, Cedar Falls, Iowa.

    Hoppe, D.M. 2005. Malformed frogs in Minnesota: history and interspecific differences. pp. 103–108 *in* Lannoo, M.J. (Ed.), Amphibian Declines: The Conservation Status of United States Species. University of California Press, Berkeley, California.

8. Volpe, E.P. and P.A. Rosenbaum. 2000. Understanding Evolution. Sixth Edition. McGraw-Hill, New York. On page 9, Hoppe's experiment and its implications are discussed.

9. Ouellet, M. 2000. Amphibian deformities: Current state of knowledge. pp. 617–661 *in* Sparling, D.W., G. Linder and C.A. Bishop (Eds.), Ecotoxicology of Amphibians and Reptiles. Society for Environmental Toxicology and Contaminants (SETAC) Press, Pensacola, Florida.

Ouellet, M., J. Bonin, J. Rodrigue, J.-L. DesGranges and S. Lair. 1997. Hindlimb deformities (ectromelia, ectrodactyly) in free-living anurans from agricultural habitats. Journal of Wildlife Diseases 33:95–104.

10. See, for example, the classic two volume set: Edeiken, J. and P.J. Hodes. 1967. Roentgen Diagnosis of Diseases of Bone. Golden's Diagnostic Radiology, Section 6. Robbins, L.L. (Ed.). Williams and Wilkins, Baltimore, Maryland.

11. Holman, J.A. 1995. Pleistocene Amphibians and Reptiles in North America. Oxford University Press, New York.

Holman, J.A. 1998. Amphibian recolonization of midwestern states in the postglacial Pleistocene. pp. 9–15 in Lannoo, M.J. (Ed.), Status and Conservation of Midwest Amphibians. University of Iowa Press, Iowa City, Iowa.

12. Labeled frog skeletons can be found at http://www.student.loretto.org/zoology/Graphic%20webs/Frog-%skeleton.htm and other internet sites.

13. Young, J.Z. 1962. The Life of Vertebrates, Second Edition, Oxford. In Porter, K.R. 1972. Herpetology. W.B. Saunders, Philadelphia, p. 38.

14. Romer, A.S. 1957. Man and the Vertebrates. University of Chicago Press, Chicago, Illinois, p. 72.

15. Handrigan, G.R. and R.J. Wassersug. 2007. The anuran Bauplan: A review of the adaptive, developmental, and genetic underpinnings of frog and tadpole morphology. Biological Reviews 82:1–25.

16. Lanyon, L.E. and C.T. Rubin. 1985. Adaptation in skeletal structures. pp. 1–25 in Hildebrand, M., D.M. Bramble, K.F. Liem and D.B. Wake (Eds.), Functional Vertebrate Morphology. Belknap Press of Harvard University Press, Cambridge, Massachusetts.

17. Bryant, S.V., D.M. Gardiner and K. Muneoka. 1987. Limb development and regeneration. American Zoologist 27:675–696.

King, M.W., T. Nguyen, J. Calley, M.W. Harty, M.C. Muzinich, A.L. Mescher, C. Chalfant, M. N'Cho, K. McLeaster, J. McEntire, D. Stocum, R.C. Smith and A.W. Neff. 2003. Identification of genes expressed during Xenopus laevis limb regeneration by using subtractive hybridization. Developmental Dynamics 226:398–409.

Moore, K.L. 1988. The Developing Human, Saunders, Philadelphia, pp. 355–363.

Muneoka, K. and S.V. Bryant. 1982. Evidence that patterning mechanisms in developing and regenerating limbs are the same. Nature 298:369–371.

Neff, A.W., M.W. King, M.W. Harty, T. Nguyen, J. Calley, R.C. Smith and A.L. Mescher. 2005. Expression of Xenopus XlSALL4 during limb development and regeneration. Developmental Dynamics 233:356–367.

Noden, D.M. and A. de Lahunta. 1985. The Embryology of Domestic Animals: Developmental Mechanisms and Malformations. Williams and Wilkins, Baltimore, Maryland, pp. 105–107.

Oppenheimer, S.B. and G. Lefevre, Jr. 1984. Introduction to Embryonic Development. Second Edition. Allyn and Bacon, Inc., Boston, Massachusetts, pp. 196–210.

Simon, H-G., R. Kittappa, P. A. Khan, C. Tsilfidis, R.A. Liversage and S. Oppenheimer. 1997. A novel family of T-box genes in urodele amphibian limb development and regeneration: candidate genes involved in vertebrate forelimb/hindlimb patterning. Development 124:1355–1366.

18. Summarized in Jacobson, M. 1991. Developmental Neurobiology. Third Edition. Plenum Press, New York.

19. Farel, P.B. 1989. Naturally-occurring cell death and differentiation of developing spinal motoneurons following axotomy. Journal of Neuroscience 9:2103–2113.

20. Arnold, S.J. and R.J. Wassersug. 1978. Differential predation on metamorphic anurans by garter snakes (Thamnophis): social behavior as a possible defense. Ecology 59:1014–1022.

Altig, R. and R.W. McDiarmid. 1999. Tadpoles: The Biology of Anuran Larvae. University of Chicago Press, Chicago, Illinois.

21. Moore, K.L. 1988. The Developing Human, Saunders, Philadelphia, pp. 170–206.

Noden, D.M. and A. de Lahunta. 1985. The Embryology of Domestic Animals: Developmental Mechanisms and Malformations. Williams and Wilkins, Baltimore, Maryland, pp. 161–168.

22. Lannoo, M.J. (Ed.) 2005. Amphibian Declines: The Conservation Status of United States Species. University of California Press, Berkeley, California.

CHAPTER 2: MALFORMED FROG TYPES

1. Freud, S. (trans. J. Riviere). 1924. Collected Writings. Volume II. Hogarth Press, London.

2. Meteyer, C.A., K.I. Loeffler, J.F. Fallon, K.A. Converse, E. Green, J.C. Helgen, S. Kersten, R. Levey, L. Eaton-Poole and J.G. Burkhart. 2000. Hind limb malformations in free-living northern leopard frogs (Rana pipiens) from Maine, Minnesota, and Vermont suggest multiple etiologies. Teratology 62:151–171.

See also: Meteyer, C.A. 2000. Field Guide to Malformations of Frogs and Toads. U.S. Geological Survey Biological Science Report USGS/BRD/BSR–

2000-0005 at http://www/nwhc.usgs.gov/publications/fact-sheets/pdfs/frog.pdf.

3. Lannoo, M.J., D.R. Sutherland, P. Jones, D. Rosenberry, R.W. Klaver, D.M. Hoppe, P.T.J. Johnson, K.B. Lunde, C. Facemire and J.M. Kapfer. 2003. Multiple causes for the malformed frog phenomenon. Pp. 233–262. *In* Linder, G., S. Krest, D. Sparling and E. Little (Eds.), Multiple Stressor Effects in Relation to Declining Amphibian Populations. American Society for Testing Materials International, West Conshohocken, Pennsylvania.

4. Drake, D.L., R. Altig, J.B. Grace and S.C. Walls. 2007. Occurrence of oral deformities in larval anurans. Copeia 2007:449–458.

5. One of the greatest gifts the scientific method has provided society is a way to properly make comparisons. The comparative method takes many forms, but is most often used when two factors vary and one wants to assess the relative value of each. For example, let's say you wish to examine the role of nature (genetics) versus nurture (environment) in human behavior. To best do this you first establish a scenario where nature is held constant but nurture varies (the ideal here is the case of identical [maternal or monozygotic] twins raised separately), and then you establish a scenario where nature varies but nurture is held constant (unrelated foster children raised in the same household). You measure the behavioral factors that most interest you and look for nature effects and nurture effects.

The great animal behaviorist Konrad Lorenz (1958) exhorted: "every time a biologist seeks to know why an organism looks and acts as it does, he must resort to the comparative method." The equally great Niko Tinbergen (1964) agreed: "[The naturalist's] main source of inspiration is comparison. Through comparison he notices both similarities between species and differences between them." As my first mentor, Dick Bovbjerg, once told me: "Make comparisons, they're always interesting."

6. See Reeves, M.K. 2006. Alaska's abnormal frogs. U.S. Fish and Wildlife Service (no number), published in April 2006, and available at http://www.fws.gov.

7. Although there is a convention for viewing human radiographs—the images are always presented as if you are looking at the patient—there is no convention for examining frog radiographs. When we take radiographs we position preserved animals so the bones are as close as possible to the film, which gives us good "focus." There is little evidence that malformations from these frogs exhibit laterality—that they occur preferentially on one side or the other (Meteyer et al.[2] show 87 left hind limb malformations, 89 right hind limb malformations; Sessions and Ruth[14] report 60 left limb malformations, 47 right limb malformations; Helgen et al. [see below] report 321 left-side malformations, 344 right-side, and 24 bilateral), therefore, for consistency of presentation and to simplify the process of making comparisons, here I show radiographs with malformations on the left side of the image.

Helgen, J.C., M.C. Gernes, S.M. Kersten, J.W. Chirhart, J.T. Canfield, D. Bowers, J. Haferman, R.G. McKinnell and D.M. Hoppe. 2000. Field investigations of malformed frogs in Minnesota 1993–97. Journal of the Iowa Academy of Science 107:96–11

8. Lannoo, M.J. 1996. Okoboji Wetlands: A Lesson in Natural History. University of Iowa Press, Iowa City, Iowa.

9. Ouellet, M., J. Bonin, J. Rodrigue, J.-L. DesGranges and S. Lair. 1997. Hindlimb deformities (ectromelia, ectrodactyly) in free-living anurans from agricultural habitats. Journal of Wildlife Diseases 33:95–104.

10. Gardiner, D.M. and D.M. Hoppe. 1999. Environmentally induced limb malformations in mink frogs (Rana septentrionalis). Journal of Experimental Zoology 284:207–216.

Takeishi, M. 1996. On the frog, Rana ornativentralis, with supernumerary limbs found at Yamada Greenery area in Kitakyushu City, Fukuoka Prefecture. Japan Bulletin of the Kitakyushu Museum of Natural History 15:119–131.

11. Summarized in: Gardiner, D.M. and D.M. Hoppe. 1999. Environmentally induced limb malformations in mink frogs (Rana septentrionalis). Journal of Experimental Zoology 284:207–216.

See also: Burke, A.C. and C. Tabin. 1996. Virally mediated misexpression of Hoxc-6 in the cervical mesoderm results in spinal nerve truncations. Developmental Biology 178:192–197.

Burke, A.C., C.E. Nelson, B.A. Morgan and C. Tabin. 1995. Hox genes and the evolution of vertebrate axial morphology. Development 121:333–346.

Chapman, D.L., N. Garvey, S. Hancock, M. Alexiou, S.I. Agulnik, J.J. Gibson-Brown, J. Cebra-Thomas, R.J. Bollag, L.M. Silver and V.E. Papaioannou. 1996. Expression of T-box family genes, Tbx1–Tbx5, during early mouse development. Developmental Dynamics 206:379–390.

Gibson-Brown, J.J., S.I. Agulnik, D.L. Chapman, M. Alexiou, N. Garvey, L.M. Silver and V.E. Papaioannou. 1996. Evidence of a role for T-box genes in the evolution of limb morphogenesis and the specification of forelimb/hindlimb identity. Mechanisms of Development 56:93–101.

Gibson-Brown, J.J., S.I. Agulnik, L.M. Silver, L. Niswander and V.E. Papaioannou. 1996. Involvement of T-box genes Tbx2–Tbx5 in vertebrate limb specification and development. Development 125:2499–2509.

Isaac, A., C. Rodriguez-Esteban, A. Ryan, M. Altabef, T. Tsukui, K. Patel, C. Tickle and J.C. Izpisaua-Belmonte. 1998. Tbx genes and limb identity in chick embryo development. Development 125:1867–1875.

Logan, M., H.G. Simon and C. Tabin. 1998. Differential regulation of T-box and homeobox transcription factors suggests roles in controlling chick limb-type identity. Development 125:2825–2835.

Oliver, G., C.V. Wright, J. Hardwicke and E.M. De Robertis. 1988. A gradient of homeodomain protein in developing forelimbs of *Xenopus* and mouse embryos. Cell 55:1017–1024.

12. Loeffler, I.K., D.L. Stocum, J.F. Fallon and C.U. Meteyer. 2001. Leaping lopsided: A review of the current hypotheses regarding etiologies of limb malformations in frogs. Anatomical Record 265:228–245.

13. Ankley, G.T., J.E. Tietge, D.L. DeFoe, K.M. Jensen, G.W. Holcombe, E.J. Durhan and S.A. Diamond. 1998. Effects of ultraviolet light and methoprene on survival and development of *Rana pipiens*. Environmental Toxicology and Chemistry 17:2530–2542.

Ankley, G.T., J.E. Tietge, G.W. Holcombe, D.L. DeFoe, S.A. Diamond, K.M. Jensen and S.J. Degitz. 2000. Effects of laboratory ultraviolet radiation and natural sunlight on survival and development of *Rana pipiens*. Canadian Journal of Zoology 78:1092–1100.

Ankley, G.T., S.A. Diamond, J.E. Tietge, G.W. Holcombe, K.M. Jensen, D.L. DeFoe and R. Peterson. 2002. Assessment of the risk of solar ultraviolet radiation to amphibians. I. Dose dependent induction of hindlimb malformations in the northern leopard frog (*Rana pipiens*). Environmental Science and Technology 36:2853–2858.

See also: Bilski, P., J.G. Burkhart and C.F. Chignell. 2003. Photochemical characterization of water samples from Minnesota and Vermont sites with malformed frogs: potential influence of photosensitization by singlet molecular oxygen and free radicals on aquatic toxicology. Aquatic Toxicology 65:229–241.

14. Sessions, S.K. and S.B. Ruth. 1990. Explanation for naturally occurring supernumerary limbs in amphibians. Journal of Experimental Zoology 254:38–47.

Sessions, S.K., R.A. Franssen and V.L. Horner. 1999. Morphological cues from multilegged frogs: Are retinoids to blame? Science 284:800–802.

15. Stopper, G.F., L. Hecker, R.A. Franssen and S.S. Sessions. 2002. How trematodes cause limb deformities in amphibians. Journal of Experimental Zoology Part B: Molecular and Developmental Evolution 294:252–263.

16. Johnson, P.T.J., K.B. Lunde, D.A. Zelmer and J.K. Werner. 2003. Limb deformities as an emerging parasitic disease in amphibians. Conservation Biology 17:1724–1737 (p. 1728).

17. Johnson, P.T.J., K.B. Lunde, E.G. Ritchie and A.E. Launer. 1999. The effect of trematode infection on amphibian limb development and survivorship. Science 284:802–804.

See also: Johnson, P.T.J., J.M. Chase, K. L. Dosch, R.B. Hartson, J.A. Gross, D.J. Larson, D.R. Sutherland and S.R. Carpenter. 2007. Aquatic eutrophication promotes pathogenic infection in amphibians. Proceedings of the National Academy of Science 104:15781–15786.

18. Bliss, M. 2005. Harvey Cushing: A Life in Surgery. Oxford, New York, p. 351.

19. Johnson, P.T.J., K.B. Lunde, D.A. Zelmer and J.K. Werner. 2003. Limb deformities as an emerging parasitic disease in amphibians. Conservation Biology 17:1724–1737.

20. Hauver, R.C. 1958. Studies on natural anomalies of the hind limbs of *Rana catesbeiana*, Master's thesis. Miami University, Oxford, Ohio.

21. Blackburn, L.M. 2001. Status of Blanchard's cricket frogs (*Acris crepitans blanchardi*) along their decline front: population parameters, malformation rates and disease. Master's thesis. Ball State University, Muncie, Indiana.

CHAPTER 3: HOTSPOTS

1. Maclean, N. 1994. Young Men and Fire. University of Chicago Press, Chicago, Illinois, p. 164.

2. Ecoregions indicated are U.S. E.P.A. Level III ecoregions as defined by the following article: United States Environmental Protection Agency (USEPA). 2000. Level III ecoregions of the continental United States (revision of Omernik, 1987, see below), Map M-1. USEPA National Health and Environmental Effects Research Laboratory, Corvallis, Oregon.

Omernik, J.M. 1987. Ecoregions of the conterminous United States. Annals of the Association of American Geographers 77:118–125.

3. Johnson, P.T.J., K.B. Lunde, D.A. Zelmer and J.K. Werner. 2003. Limb deformities as an emerging parasitic disease in amphibians: Evidence from museum specimens and re-survey data. Conservation Biology 17:1724–1737.

4. Lannoo, M.J., D.R. Sutherland, P. Jones, D. Rosenberry, R.W. Klaver, D.M. Hoppe, P.T.J. Johnson, K.B. Lunde, C. Facemire and J.M. Kapfer. 2003. Multiple causes for the malformed frog phenomenon. Pp. 233–262. *In* Linder, G., S. Krest, D. Sparling and E. Little (Eds.), Multiple Stressor Effects in Relation to Declining Amphibian Populations. American Society for Testing Materials International, West Conshoshocken, Pennsylvania.

5. Among my professional friends, there are few people I have known longer than Dan Sutherland. We met in 1977, during my first year at the Iowa Lakeside Laboratory. Dan was a Ph.D student working under the icon, Dr. Martin Ulmer, who was a former President of the American Society of Parasitologists and a true free spirit. Dan and I shared an affliction for volleyball, and for several years we played every non-stormy evening at the Lab. When we (and our group of like-minded jocks) returned to Iowa State's campus each fall, we did well in the

intramural leagues, which at that time included some pretty hotshot fraternity teams. We won a drawer-full of championship t-shirts, which I wore well into the 1980s, when they took on a second life as shop rags. When Dan and I were asked to work together in 2001, it had the feel of old Poncho and Cisco strapping on their revolvers for one more go. We borrowed an ancient U.S.G.S. pickup truck from the Upper Midwest Environmental Sciences Center in LaCrosse. Neither the air conditioning nor the windows worked, which just about killed us when we began sampling and temperatures were near 100°F (we hydrated on Gatorade and beer). We learned to appreciate the heat two weeks later, when nighttime temperatures went below freezing (we substituted hot chocolate for Gatorade). For that time we worked together, it was like playing volleyball again, without the t-shirt trophies.

Dan was only 54 when he died on May 26, 2006. He was the one classically trained parasitologist working on amphibian malformations. Because of the rigor of his training and his own personal drive to get the story right, Dan was also the lone diagnostician who felt that *Ribeiroia* metacercariae could be definitively identified only in freshly dissected tissue. He had noted that when vertebrate tissues are preserved, too much happens to embedded trematode metacercariae to allow confidence, and he mistrusted every paper or report where identifications were based on fixed tissues. Dan felt that the artifacts created too much opportunity for biologists to see what they wanted to see instead of what was there. I was at Lakeside when Dan was being trained; I saw what he went through. If Dan said a work was not up to standard—and he said this more than once about parasite/malformation papers—I could not disagree. When we were on our "expedition," Dan was dissecting parasites from northern leopard frogs collected from an Iowa site and he said: "These animals have been metamorphosed for a long time." I asked him how he knew and he showed me a lung parasite in the genus *Hematoloechus* sp. that frogs acquire from ingesting infected dragonflies (Sutherland and Kapfer, unpublished data). These frogs had to have been on land long enough to feed on dragonflies, transmit the parasite, and then have the parasite develop in the frog. I remember at the time thinking that was about the neatest thing I had ever heard, and how differently Dan and I saw the world. His death was a huge personal loss for me, and removed some necessary depth of understanding, and therefore rigor, from the discipline; this may delay a true understanding of the role that parasites play in causing amphibian malformations.

6. Hoppe, D.M. 2002. Mortality and population declines associated with a Minnesota malformed frog site. Pp. 77–85. *In* McKinnell, R.G. and D.L. Carlson (Eds.), Proceedings of the Sixth International Symposium on the Pathology of Reptiles and Amphibians. University of Minnesota, St. Paul, Minnesota.

7. NARCAM (North American Reporting Center for Amphibian Malformations) data set at: www.frogweb.nbii.gov/narcam. Accessed on 16 October 2006.

8. In this paper, Johnson and his colleagues examine records from nine historical sites and claim to have isolated *Ribeiroia* metacercariae from museum specimens at six of these sites. One point of contention involved the techniques used to identify trematode metacercariae. In their own words (pp. 1727–1728): "To determine the parasite species present at each locality, we identified metacercariae by direct observation of diagnostic characteristics, such as the esophageal diverticula found in *Ribeiroia* specimens. In some cases direct identification was precluded by the condition or, less frequently, orientation of the metacercariae. Thus, we used parametric discriminant-function analysis to predict the group membership of all metacercariae." That is, where diagnostic morphological features could not be used to determine identity, Johnson and his colleagues used "taxonomy by shape." Dan Sutherland and I had a long discussion about this. The bottom line is, in gathering historical data on parasite loads, there will always be some compromises involving assumptions and guesswork. Different researchers draw different lines regarding what are acceptable data and what are not.

9. Converse, K.A., J. Mattsson and L. Eaton-Poole. 2000. Field surveys of Midwestern and Northeastern Fish and Wildlife Service lands for the presence of abnormal frogs and toads. Journal of the Iowa Academy of Science 107:160–167.

10. Reeves, M.K. 2006. Alaska's abnormal frogs. U.S. Fish and Wildlife Service (no number), published in April, 2006 and available at http://www.fws.gov.

11. Guderyahn, L. 2006. Nationwide assessment of morphological abnormalities observed in amphibians collected from United States National Wildlife Refuges. Masters Thesis, Biology Department, Ball State University, Muncie, Indiana.

CHAPTER 4: CAUSES

1. Hamlet, Prince of Denmark. Polonius act II, scene ii, lines 108–112.

2. Ouellet, M. 2000. Amphibian deformities: Current state of knowledge. Pp. 617–661. *In* Sparling, D.W., G. Linder and C.A. Bishop (Eds.), Ecotoxicology of Amphibians and Reptiles. Society for Environmental Toxicology and Contaminants (SETAC) Press, Pensacola, Florida.

 See also: Ouellet, M., J. Bonin, J. Rodrigue, J.-L. DesGranges and S. Lair. 1997. Hindlimb deformities (ectromelia, ectrodactyly) in free-living anurans from agricultural habitats. Journal of Wildlife Diseases 33:95–104.

3. Johnson, P.T.J., K.B. Lunde, E.G. Ritchie, J.K. Reaser and A.E. Launer. 2001. Morphological abnormality patterns in a California amphibian community. Herpetologica 57:336–352.

4. With the exception of Martin Ouellet, no North American scientist cites the important work of the Frenchman Jean Rostand. In addition to his books and papers on malformed amphibians (see selected references immediately below) he is the subject of a biography (by Tetry, also immediately below). His work has been ignored in part because it was written in French, because the malformations he observes do not match those in North America, because he cites different causes for these malformations, and because people tend to be unaware of older literature.

Rostand, J. 1955. Les Crapauds, les Grenouilles, et Quelques Grands Problèmes Biologiques. Gallimard, Paris.

Rostand, J. 1971. Les Étangs à Monstres, Histoire d'une Recherché (1947–1970). Stock, Paris.

Tétry, A. 1983. Jean Rostand: Proféte Clairvoyant et Fraternal. Gallimard, Paris.

5. Adams, M.J., D.E. Schindler and R.B. Bury. 2001. Association of amphibians with attenuation of ultraviolet-B radiation in montane ponds. Oecologia 128:519–525.

Anzalone, C.R., L.B. Kats and M.S. Gordon. 1998. Effects of solar UV-B radiation on embryonic development in Hyla cadaverina, Hyla regilla, and Taricha torosa. Conservation Biology 12:646–653.

Blaustein, A.R., P.D Hoffman, D.G. Hokit, J.M. Kiesecker, S.C. Walls and J.B. Hays. 1994. UV repair and resistance to solar UV-B in amphibian eggs: A link to population declines? Proceedings of the National Academy of Sciences 91:1791–1795.

Blaustein, A.R., P.D Hoffman, J.M. Kiesecker and J.B. Hays. 1994. DNA repair activity and resistance to solar UV-B radiation in eggs of the red-legged frog. Conservation Biology 10:1398–1402.

Blaustein, A.R., J.M. Kiesecker, D.P. Chivers and R.G. Anthony. 1997. Ambient UV-B radiation causes deformities in amphibian embryos. Proceedings of the National Academy of Sciences 94:13735–13737.

Corn, P.S. 1998. Effects of ultraviolet radiation on boreal toads in Colorado. Ecological Applications 8:18–26.

Crump, D., M. Berrill, D. Coulson, D. Lean, L. McGillivray and A. Smith. 1999. Sensitivity of amphibian embryos, tadpoles, and larvae to enhanced UV-B radiation in natural pond conditions. Canadian Journal of Zoology 77:1956–1966.

Hays, J.B., A.R. Blaustein, J.M. Kiesecker, P.D. Hoffman, I. Pandelova, D. Coyle and T. Richardson. 1996. Developmental responses of amphibians to solar and artificial UVB sources: a comparative study. Photochemistry and Photobiology 64:449–456.

See also: Ovaska, K., T.M. Davis and I.N. Flamarique. 1997. Hatching success and larval survival of the frogs *Hyla regilla* and *Rana aurora* under ambient and artificially enhanced solar ultraviolet radiation. Canadian Journal of Zoology 75:1081–1088.

Smith, G.R., W.A. Waters and J.E. Rettig. 2000. Consequences of embryonic UV-B exposure on the growth and development of plains leopard frog tadpoles (*Rana blairi*). Conservation Biology 14:1903–1907.

Tietge, J.E., S.A. Diamond, G.T. Ankley, D.L. DeFoe, G.W. Holcombe, K.M. Jensen, S.J. Degitz, G.E. Elonen and E. Hammer. 2001. Ambient solar UV radiation causes mortality in larvae of three species of *Rana* under controlled exposure conditions. Photochemistry and Photobiology 74:261–268.

Worrest, R.C. and D.J. Kimeldorf. 1975. Photoreactivation of potentially lethal, UV-induced damage to boreal toad (*Bufo boreas boreas*) tadpoles. Life Sciences 17:1545–1550.

6. Ankely, G.T., J.E. Tietge, D.L. DeFoe, K.M. Jensen, G.W. Holcombe, E.J. Durhan and S.A. Diamond. 1998. Effects of ultraviolet light and methoprene on survival and development of *Rana pipiens*. Environmental Toxicology and Chemistry 17:2530–2542.

Ankley, G.T., J.E. Tietge, G.W. Holcombe, D.L. DeFoe, S.A. Diamond, K.M. Jensen and S.J. Degitz. 2000. Effects of laboratory ultraviolet radiation and natural sunlight on survival and development of *Rana pipiens*. Canadian Journal of Zoology 78:1092–1100.

Ankley, G.T., S.A. Diamond, J.E. Tietge, G.W. Holcombe, K.M. Jensen, D.L. DeFoe and R. Peterson. 2002. Assessment of the risk of solar ultraviolet radiation to amphibians. I. Dose-dependent induction of hindlimb malformations in the northern leopard frog (*Rana pipiens*). Environmental Science and Technology 36:2853–2858.

7. See also: Skelly, D.K., S.R. Bolden, M.P. Holland, L.K. Freidenburg, N.A. Freidenfelds and T.R. Malcolm. 2006. Urbanization and disease in amphibians. Pp. 153–167. *In* Collinge, S. and C. Ray (Eds.), Ecology of Disease: Community Context and Pathogen Dynamics. Oxford University Press, Oxford, England.

Also: Forson, D.D. and A. Storfer. 2006. Atrazine increases ranavirus susceptibility in the tiger salamander, *Ambystoma tigrinum*. Ecological Applications 16:2325–2332.

8. Johnson, P.T.J., K.B. Lunde, E.G. Ritchie and A.E. Launer. 1999. The effect of trematode infection on amphibian limb development and survivorship. Science 284:802–804.

Johnson, P.T.J., K.B. Lunde, R.W. Haight, J. Bowerman and A.R. Blaustein. 2001. *Ribeiroia ondatrae* (Trematoda: Digena) infection induces severe limb malformations in western toads (*Bufo boreas*). Canadian Journal of Zoology 79:370–379.

Johnson, P.T.J., K.B. Lunde, E.G. Ritchie, J.K. Reaser and A.E. Launer. 2001. Morphological abnormality patterns in a California amphibian community. Herpetologica 57:336–352.

Johnson, P.T.J., K.B. Lunde, E.M. Thurman, E.G. Ritchie, S.N. Wray, D.R. Sutherland, J.M. Kapfer, T.J. Frest, J. Bowerman and A.R. Blaustein. 2002. Parasite (*Ribeiroia ondatrae*) infection linked to amphibian malformations in the western United States. Ecological Monographs 72:151–168.

Johnson, P.T.J. and K.B. Lunde. 2005. Parasite infection and limb malformations: a growing problem in amphibian conservation. Pp. 124–138. *In* Lannoo, M.J. (Ed.), Amphibian Declines: The Conservation Status of United States Species. University of California Press, Berkeley, California.

See also: Johnson, P.T.J., J.M. Chase, K. L. Dosch, R.B. Hartson, J.A. Gross, D.J. Larson, D.R. Sutherland and S.R. Carpenter. 2007. Aquatic eutrophication promotes pathogenic infection in amphibians. Proceedings of the National Academy of Science 104:15781–15786.

9. Hayes, T.B., P. Case, S. Chui, D. Chung, C. Haefele, K. Haston, M. Lee, V.P. Mai, Y. Marjuoa, J. Parker and M. Tsui. 2006. Pesticide mixtures, endocrine disruption, and amphibian declines: Are we underestimating the impact? Environmental Health Perspectives 114:(available online at <http://dx.doi.org/> use doi:10.1289/ehp.8051).

Forson, D.D. and A. Storfer. 2006. Atrazine increases ranavirus susceptibility in the tiger salamander, *Ambystoma tigrinum*. Ecological Applications 16:2325–2332.

Harris, M.L., L. Chora, C.A. Bishop and J.P. Bogart. 2000. Species- and age-related differences in susceptibility to pesticide exposure for two amphibians, *Rana pipiens* and *Bufo americanus*. Bulletin of Environmental Contamination and Toxicology 64:263–270.

Reylea, R.A. 2005. The impact of insecticides and herbicides on the biodiversity and productivity of aquatic communities. Ecological Applications 15:618–627.

Reylea, R.A. 2005. The lethal impact of Roundup and predatory stress on six species of North American tadpoles. Archives of Environmental Contamination and Toxicology 48:351–357.

Reylea, R.A. 2005. The lethal impact of Roundup on aquatic and terrestrial amphibians. Ecological Applications 15:1118–1124.

Reylea, R.A. and N. Mills. 2001. Predator-induced stress makes the pesticide carbaryl more deadly to gray treefrog tadpoles (*Hyla versicolor*). Proceedings of the National Academy of Sciences 98:2491–2496.

Reylea, R.A., N.A. Schoeppner and J.T. Hoverman. 2005. Pesticides and amphibians: The importance of community context. Ecological Applications 15:1125–1134.

Semlitsch, R.D. and C.M. Bridges. 2005. Amphibian ecotoxicology. Pp. 241–243. *In* Lannoo, M.J. (Ed.), Amphibian Declines: The Conservation Status of United States Species. University of California Press, Berkeley, California.

10. Christin, M-S., A.D. Gendron, P. Brousseau, L. Menard, D.J. Marcogliese, D. Cyr, S. Ruby and M. Fournier. 2003. Effects of agricultural pesticides on the immune system of *Rana pipiens* and on its resistance to parasitic infection. Environmental Toxicology and Chemistry 22:1127–1133.

11. Kiesecker, J.M. 2002. Synergism between trematode infection and pesticide exposure: a link to amphibian limb deformities in nature? Proceedings of the National Academy of Sciences 99:9900–9904.

12. Sutherland, D. 2005. Parasites of North American frogs. Pp. 109–123. *In* Lannoo, M.J. (Ed.), Amphibian Declines: The Conservation Status of North American Species. University of California Press, Berkeley, California.

13. Lafferty, K.D. and A.M. Kuris. 1999. How environmental stress affects the impacts of parasites. Limnology and Oceanography 44:925–931.

Lafferty, K.D. and R.D. Holt. 2003. How should environmental stress affect the population dynamics of disease? Ecology Letters 6:654–664.

14. Thiemann, G.W. and R.J. Wassersug. 2000. Patterns and consequences of behavioral responses to predators and parasites in *Rana* tadpoles. Biological Journal of the Linnean Society 71: 513–528.

See also: Taylor, C.N., K.L. Oseen and R.J. Wassersug. 2004. On the behavioral response of *Rana* and *Bufo* tadpoles to echinostomatid cercariae: Implications to synergistic factors influencing trematode infections in anurans. Canadian Journal of Zoology 82:701–706.

15. Bridges, C.M. 1997. Tadpole swimming performance and activity affected by acute exposure to sublethal levels of carbaryl. Environmental Toxicology and Chemistry 16:1935–1939.

Bridges, C.M. 1999. Effects of a pesticide on tadpole activity and predator avoidance behavior. Journal of Herpetology 33:303–306.

Bridges, C.M. 2000. Long-term effects of pesticide exposure at various life stages of the southern leopard frog (*Rana sphenocephala*). Archives for Environmental Contamination and Toxicology 39:91–96.

16. Lee, B.G., S.B. Griscom, J.S. Lee, J.C. Heesun, C.H. Koh, S.N. Luoma and N.S. Fisher. 2000. Influences of dietary uptake and reactive sulfides on metal bioavailability from aquatic sediments. Science 287:282–284.

Sparling, D.W. and T.P. Lowe. 1996. Metal concentrations of tadpoles in experimental ponds. Environmental Pollution 91:149–159.

17. Gardiner, D.M. and D.M. Hoppe. 1999. Environmentally induced limb malformations in mink frogs (*Rana septentrionalis*). Journal of Experimental Zoology 284:207–216.

18. Johnson, P.T.J., K.B. Lunde, E.G. Ritchie and A.E. Launer. 1999. The effect of trematode infection on amphibian limb development and survivorship. Science 284:802–804.

CHAPTER 5: RESOLUTIONS

1. Cohen, L. 1992. Anthem. Stranger Music Incorporated, New York.

2. Blaustein, A.R. and P.T.J. Johnson. 2003. Explaining frog deformities. Scientific American, February:60–65.

3. Stocum, D.L. 2000. Frog limb deformities: an "eco-devo" riddle wrapped in multiple hypotheses surrounded by insufficient data. Teratology 62:147–150.

4. Souder, W. 2000. A Plague of Frogs: The Horrifying True Story. Hyperion Press, New York.

5. Meteyer, C.A., K.I. Loeffler, J.F. Fallon, K.A. Converse, E. Green, J.C. Helgen, S. Kersten, R. Levey, L. Eaton-Poole and J.G. Burkhart. 2000. Hind limb malformations in free-living northern leopard frogs (*Rana pipiens*) from Maine, Minnesota, and Vermont suggest multiple etiologies. Teratology 62:151–171.

6. Steinbeck, J. 1951. The Log from the Sea of Cortez: About Ed Ricketts. Viking Press, New York.

7. Darwin, C. 1859. On the Origin of Species by Means of Natural Selection or The Preservation of Favoured Races in the Struggle for Life. John Murray, Albemarle Street, London, England.

8. Except Georges-Louis Leclerc, Comte de Buffon (1707–1788). Steven T. Asma (2001. Stuffed Animals and Pickled Heads: The Culture and Evolution of Natural History Museums. Oxford University Press, Oxford) writes (pp. 123–124): "For the system lovers, nature came packaged in neat rational bundles or norms, and variations from those norms were thought to be rare freaks of little consequence. Animal and plant variations were just occasional obstacles to the otherwise smooth process of classification, random square pegs that did not fit nicely into the round holes. For Buffon, on the other hand, nature was variation through and through. There were no truly round pegs. Taxonomists simply ignored the subtle variations and forced different-shaped pegs into holes of their own invention. Buffon says of nature, 'It is necessary to see nothing as impossible, to expect anything, and to suppose that all that can be is. Ambiguous species, irregular productions, anomalous beings will from now on cease to astonish us.'"

9. Windsor, M.P. 1991. Reading the Shape of Nature. University of Chicago Press, Chicago, Illinois.

10. Gould, S.J. and R.C. Lewontin. 1979. The spandrels of San Marco and the Panglossian paradigm: a critique of the adaptationist programme. Proceedings of the Royal Society of London, Series B. 205:581–598.

11. Stopper, G.F., L. Hacker, R.A. Franssen and S.K. Sessions. 2002. How trematodes cause limb deformities in amphibians. Journal of Experimental Zoology Part B: Molecular and Developmental Evolution 294:252–263.

12. Carroll, R.L. 1988. Vertebrate Paleontology and Evolution. W.H. Freeman and Company, New York.

   McDiarmid, R.W. and R. Altig. 1999. Introduction: The tadpole arena. Pp. 1–23. *In* McDiarmid, R.W. and R. Altig (Eds.), Tadpoles: The Biology of Anuran Larvae. University of Chicago Press, Chicago, Illinois.

13. Handrigan, G.R. and R.J. Wassersug. 2007. The anuran *Bauplan*: A review of the adaptive, developmental, and genetic underpinnings of frog and tadpole morphology. Biological Reviews 82:1–25.

14. Merrell, D.J. 1969. Natural selection in a leopard frog population. Journal of the Minnesota Academy of Science 35:86–89.

15. Helgen, J.C., M.C. Gernes, S.M. Kersten, J.W. Chirhart, J.T. Canfield, D. Bowers, J. Haferman, R.G. McKinnell and D.M. Hoppe. 2000. Field investigations of malformed frogs in Minnesota 1993–97. Journal of the Iowa Academy of Science 107:96–112.

16. Johnson, P.T.J., K.B. Lunde, E.G. Ritchie, J.K. Reaser and A.R. Launer. 2001. Morphological abnormality patterns in a California amphibian community. Herpetologica 57:336–352.

17. Sessions, S.K. and S.B. Ruth. 1990. Explanation for naturally occurring supernumerary limbs in amphibians. Journal of Experimental Zoology 254:38–47.

18. Helgen, J., R.G. McKinnell and M.C. Gernes. 1998. Investigation of malformed northern leopard frogs in Minnesota. Pp. 288–297. *In* Lannoo, M.J. (Ed.), Status and Conservation of Midwestern Amphibians. University of Iowa Press, Iowa City, Iowa.

19. Schotthoefer, A.M., A.V. Koehler, C.U. Meteyer and R.A. Cole. 2003. Influence of *Ribeiroia ondatrae* (Trematoda: Digenea) infection on limb development and survival of northern leopard frogs (*Rana pipiens*): effects of host stage and parasite-exposure level. Canadian Journal of Zoology 81:1144–1153.

20. Johnson, P.T.J., K.B. Lunde, E.M. Thurman, E.G. Ritchie, S.N. Wray, D.R. Sutherland, J.M. Kapfer, T.J. Frest, J. Bowerman and A.R. Blaustein. 2002.

Parasite (*Ribeiroia ondatrae*) infection linked to amphibian malformations in the western United States. Ecological Monographs 72:151–168.

21. Gardiner, D., A. Ndayibagira, F. Grün and B. Blumberg. 2003. Topic 4.6: Deformed frogs and environmental retinoids. Pure and Applied Chemistry 75:2263–2273.

22. Burnham, K.B. and D.R. Anderson. 2002. Model Selection and Multimodal Inference: A Practical Information-Theoretic Approach. Second Edition. Springer-Verlag, New York.

23. Asma, S.T. 2001. Stuffed Animals and Pickled Heads: The Culture and Evolution of Natural History Museums. Oxford University Press, Oxford, England.

24. Maclean, N. 1994. Young Men and Fire. University of Chicago Press, Chicago, Illinois, p. 164.

25. NARCAM (North American Reporting Center for Amphibian Malformations) dataset at: www.frogweb.nbii.gov/narcam. Accessed on 16 October 2006.

26. For information on the home ranges of all United States and Canadian amphibian species see the accounts in Lannoo, M.J. (Ed.) 2005. Amphibian Declines: The Conservation Status of United States Species. University of California Press, Berkeley, California.

27. Ouellet, M. 2000. Amphibian deformities: Current state of knowledge. Pp. 617–661. *In* Sparling, D.W., G. Linder and C.A. Bishop (Eds.), Ecotoxicology of Amphibians and Reptiles. Society for Environmental Toxicology and Contaminants (SETAC) Press, Pensacola, Florida.

28. Blaustein, A.R. and P.T.J. Johnson. 2003. The complexity of deformed amphibians. Frontiers in Ecology and the Environment 1:87–94.

29. Hoppe, D.M. 2005. Malformed frogs in Minnesota: History and interspecific differences. Pp. 103–108. *In* Lannoo, M.J. (Ed.), Amphibian Declines: The Conservation Status of United States Species. University of California Press, Berkeley, California.

30. Johnson, P.T.J., K.B. Lunde, R.W. Haight, J. Bowerman and A.R. Blaustein. 2001. *Ribeiroia ondatrae* (Trematoda: Digenea) infection induces severe limb malformations in western toads (*Bufo boreas*). Canadian Journal of Zoology 79:370–379.

31. Johnson, P.T.J., K.B. Lunde, D.A. Zelmer and J.K. Werner. 2003. Limb deformities as an emerging parasitic disease in amphibians: Evidence from museum specimens and re-survey data. Conservation Biology 17:1724–1737.

32. Prine, J. 1975. Common Sense. Atlantic Records, New York.

33. Lannoo, M.J., D.R. Sutherland, P. Jones, D. Rosenberry, R.W. Klaver, D.M. Hoppe, P.T.J. Johnson, K.B. Lunde, C. Facemire and J.M. Kapfer. 2003. Multiple causes for the malformed frog phenomenon. Pp. 233–262. *In* Linder, G.,

S. Krest, D. Sparling and E. Little (Eds.), Multiple Stressor Effects in Relation to Declining Amphibian Populations. American Society for Testing Materials International, West Conshoshocken, Pennsylvania.

34. Johnson, P.T.J., K.B. Lunde, E.G. Ritchie and A.E. Launer. 1999. The effect of trematode infection on amphibian limb development and survivorship. Science 284:802–804.

See also: Johnson, P.T.J., J.M. Chase, K. L. Dosch, R.B. Hartson, J.A. Gross, D.J. Larson, D.R. Sutherland and S.R. Carpenter. 2007. Aquatic eutrophication promotes pathogenic infection in amphibians. Proceedings of the National Academy of Science 104:15781–15786.

35. Kiesecker, J. 2002. Synergism between trematode infection and pesticide exposure: A link to amphibian limb deformities in nature? Proceedings of the National Academy of Science 99:9900–9904.

36. Johnson, P.T.J., D.R. Sutherland, J.M. Kinsella and K.B. Lunde. 2004. Review of the trematode genus *Ribeiroia* (Psilistomatidae): Ecology, life history and pathogenesis with special emphasis on the amphibian malformation problem. Advances in Parasitology 57:191–253.

37. Eaton, B.R., S. Eaves, C. Stevens, A. Puchniak and C.A. Paszkowski. 2004. Deformity levels in wild populations of the wood frog (*Rana sylvatica*) in three ecoregions of western Canada. Journal of Herpetology 38: 283–287.

38. Bonin, J., M. Ouellet, J. Rodrigue, J.-L. DesGranges, F. Gagné, T.F. Sharbel and L.A. Lowcock. 1997. Measuring the health of frogs in agricultural habitats subjected to pesticides. Pp. 258–270. *In* Green, D.M. (Ed.), Amphibians in Decline: Canadian Studies of a Global Problem. Herpetological Conservation, Number 1, Society for the Study of Amphibians and Reptiles, St. Louis, Missouri.

Lowcock, L.A., T.F. Sharbel, J. Bonin, M. Ouellet, J. Rodrigue and J.-L. DesGranges. 1997. Flow cytometric assay for in vivo genotoxic effects of pesticides in green frogs (*Rana clamitans*). Aquatic Toxicology 38:241–255.

Ouellet, M., J. Bonin, J. Rodrigue, J.-L. DesGranges and S. Lair. 1997. Hindlimb deformities (ectromelia, ectrodactyly) in free-living anurans from agricultural habitats. Journal of Wildlife Diseases 33:95–104.

39. Taylor, B., D. Skelly, L.K. Demarchis, M.D. Slade, D. Galusha and P.M. Rabinowitz. 2005. Proximity to pollution sources and risk of amphibian limb malformation. Environmental Health Perspectives 113:1497–1501.

40. Piha, H., M. Pekkonen and J. Merilä. 2006. Morphological abnormalities in amphibians in agricultural habitats: A case study of the common frog *Rana temporaria*. Copeia 2006:810–817.

41. Gurushankara, H.P., S.V. Krishnamurthy and V. Vasudev. 2007. Morphological abnormalities in natural populations of common frogs inhabiting agro-ecosystems of central Western Ghats. Applied Herpetology 4:39–45.

42. Lannoo, M.J. Unpublished data.

43. Gardiner, D.M. and D.M. Hoppe. 1999. Environmentally induced limb mal-formations in mink frogs (*Rana septentrionalis*). Journal of Experimental Zoology 284:207–216.

44. Levey, R. 2003. Investigations into the causes of amphibian malformations in the Lake Champlain Basin of New England. Final Report to the Vermont Department of Environmental Conservation, Waterbury, Vermont.

45. Blaustein, A.R., J.M. Kiesecker, D.P. Chivers and R.G. Anthony. 1997. Ambient UV-B radiation causes deformities in amphibian embryos. Proceedings of the National Academy of Sciences 94:13735–13737.

46. Ankely, G.T., J.E. Tietge, D.L. DeFoe, K.M. Jensen, G.W. Holcombe, E.J. Durhan and S.A. Diamond. 1998. Effects of ultraviolet light and methoprene on survival and development of *Rana pipiens*. Environmental Toxicology and Chemistry 17:2530–2542.

47. La Clair, J.J., J.A. Bantle and J. Dumont. 1998. Photoproducts and metabo-lites of a common insect growth regulator produce developmental deformities in *Xenopus*. Environmental Science and Technology 32:1453–1461.

48. Fernandez, M. and J. l'Haridan. 1992. Influence of lighting conditions on toxicity and genotoxicity of various PAH in the newt in vivo. Mutation Research 298:31–41.

See also: Hatch, A.C and G.A. Burton Jr. 1998. Effects of photoinduced toxicity of fluoranthene on amphibian ambryos and larvae. Environmental Toxicology and Chemistry 17:1777–1785.

49. http://www.panna.org/resources/panups/panup_20060228.dv.html

50. Degitz, S.J., G.W. Holcombe, P.A. Kosian, J.E. Tietge, E.J. Durhan and G.T. Ankley. 2003. Comparing the effects of stage and duration of retinoic acid exposure on amphibian limb development: Chronic exposure results in mortal-ity, not limb malformations. Toxicological Science 74:139–146.

51. Degitz, S.J., P.A. Kosian, E.A. Makynen, K.M. Jensen and G.T. Ankley. 2000. Stage- and species-specific developmental toxicity of all-trans retinoic acid in four native north American ranids and *Xenopus laevis*. Toxicological Sciences 57:264–274.

52. Wassersug, R.J. 1997. Where the tadpole meets the world: Observations and speculations on biomechanical and biochemical factors that influence metamorpho-sis in anurans. American Zoologist 37:124–136. In this paper, Wassersug states (pp. 129–130): " . . . there is no reason to believe that the gene regulation program for any particular larval features in [*Xenopus*] is the same in other anuran genera."

53. Carson, R. 1963. Silent Spring. Houghton Mifflin, Boston, Massachusetts.

54. Hayes, T., K. Haston, M. Tsui, A. Hoang, C. Haeffle and A. Vonk. 2002. Feminization of male frogs in the wild. Water-borne herbicide threatens amphibian populations in parts of the United States. Nature 419:895–896.

Hayes, T.B., A. Collins, M. Lee, M. Mendoza, N. Noriegs, A.A. Stuart and A. Vonk. 2002. Hermaphroditic, demasculinized frogs after exposure to the herbicide atrazine at low ecologically relevant doses. Proceedings of the National Academy of Sciences 99:5476–5480.

Hayes, T., K. Haston, M. Tsui, A. Hoang, C. Haeffle and A. Vonk. 2003. Atrazine-induced hermaphroditism at 0.1 ppb in American leopard frogs (*Rana pipiens*): Laboratory and field evidence. Environmental Health Perspectives 111:568–575.

55. Tietge, J.E., G.T. Ankley, D.L. DeFoe, G.W. Holcombe and K.M. Jensen. 2000. Effects of water quality on development of *Xenopus laevis*: A frog embryo teratogenesis assay–*Xenopus* assessment of surface water associated with malformations in native anurans. Environmental Toxicology and Chemistry 19:2114–2121.

56. Burkhart, J.G., J.C. Helgen, D.J. Fort, K. Gallagher, D. Bowers, T.L. Propst, M. Gernes, J. Magner, M.D. Shelby and G. Lucier. 1998. Induction of mortality and malformation in *Xenopus laevis* embryos by water sources associated with field frog deformities. Environmental Health Perspectives 106:841–848.

57. Fort, D.J., T.L Propst, E.L. Stover, J.C. Helgen, R.B. Levey, K. Gallagher and J.G. Burkhart. 1999. Effects of pond water, sediment, and sediment extracts from Minnesota and Vermont, USA, on early development and metamorphosis of *Xenopus*. Environmental Toxicology and Chemistry 18:2305–2315.

58. Fort, D.J., R.L. Rogers, H.F. Copley, L.A. Bruning, E.L. Stover, J.C. Helgen and J.G. Burkhart. 1999. Progress toward identifying causes of maldevelopment induced in *Xenopus* by pond water and sediment extracts from Minnesota, USA. Environmental Toxicology and Chemistry 18:2316–2324.

59. Hopkins, W.A., J. Congdon and J.K. Ray. 2000. Incidence and impact of axial malformations in larval bullfrogs (*Rana catesbeiana*) developing in sites polluted by a coal-burning power plant. Environmental Toxicology and Chemistry 19:862–868.

60. Rowe, C.L., O.M. Kinney and J.D. Congdon. 1998. Oral deformities in tadpoles of the bullfrog (*Rana catesbeiana*) caused by conditions in a polluted habitat. Copeia 1998:244–246.

61. Bridges, C.M. 2000. Long-term effects of pesticide exposure at various life stages of the southern leopard frog (*Rana sphenocephala*). Archives for Environmental Contamination and Toxicology 39:91–96.

Bridges, C.M. and R.D. Semlitsch. 2005. Xenobiotics. Pp. 89–92. *In* Lannoo, M.J. (Ed.), Amphibian Declines: The Conservation Status of United States Species. University of California Press, Berkeley, California.

62. Sparling, D.W., G. Linder and C.A. Bishop. 2000. Ecotoxicology of Amphibians and Reptiles. Society for Environmental Toxicology and Contaminants (SETAC) Press, Pensacola, Florida.

63. Cowman, D.F. and L.E. Mazanti. 2000. Ecotoxicology of "new generation" pesticides to amphibians. Pp. 233–268. *In* Sparling, D.W., G. Linder and C.A. Bishop (Eds.), Ecotoxicology of Amphibians and Reptiles. Society for Environmental Toxicology and Contaminants (SETAC) Press, Pensacola, Florida.

64. Geen, G.H., B.A. McKeown, T.A. Watson and D.B. Parker. 1984. Effects of acephate (orthene) on development and survival of the salamander, *Ambystoma gracile* (Baird). Journal of Environmental Science and Health B19:157–170.

65. Pandey, A.K. and V. Tomar. 1985. Melanophores in *Bufo melanosticus* (Schneider) tadpoles following exposure to the insecticide dimethoate. Bulletin of Environmental Contamination and Toxicology 35:796–801.

66. Tomar,V. and A.K. Pandey. 1988. Sublethal toxicity of an insecticide to the epidermal melanophores of *Rana* tadpoles. Bulletin of Environmental Contamination and Toxicology 41:582–588.

67. Snawder, J.E. and J.E. Chambers. 1989. Toxic and developmental effects of organophosphorous insecticides in embryos of the South African clawed frog. Journal of Environmental Science and Health B24:205–218.

Snawder, J.E. and J.E. Chambers. 1990. Critical time periods and the effect of tryptophan in malathion-induced developmental defects in *Xenopus* embryos. Life Science 46:1635–1642.

68. Honrubia, M.P., M.P. Herraez and R. Alvarez. 1993. The carbamate insecticide ZZ-Aphox induced structural changes of gills, liver, gall-bladder, heart, and notochord of *Rana perezi* tadpoles. Archives of Environmental Contamination and Toxicology 25:184–191.

Alvarez, R., M.P. Honrubia and M.P. Herráez. 1995. Skeletal malformations induced by the insecticides ZZ-Aphox® and Folidol® during larval development of *Rana perezi*. Archives of Environmental Contamination and Toxicology 28:349–356.

69. Fulton, M.H. and J.E. Chambers. 1985. The toxic and teratogenic effects of selected organophosphorous compounds on the embryos of three species of amphibians. Toxicology Letters 26:175–180.

70. Harris, M.L., C.A. Bishop, J. Struger, B. Ripley and J.P. Bogart. 1998. The functional integrity of northern leopard frog (*Rana pipiens*) and green frog (*Rana clamitans*) populations in orchard wetlands. II. Effects of pesticides and eutrophic conditions on early life stage development. Environmental Toxicology and Chemistry 17:1351–1363.

71. Pawar, K.R. and M. Katdare. 1984. Toxic and teratogenic effects of feni-trothion, BHC and carbofuran on embryonic development of the frog *Microhyla ornata*. Toxicology Letters 22:7–13.

72. Lowcock, L.A., T.F. Sharbel, J. Bonin, M. Ouellet, J. Rodrigue and J-L. DesGranges. 1997. Flow cytometric assay for in vivo genotoxic effects of pesti-cides in green frogs (*Rana clamitans*). Aquatic Toxicology 38:241–255.

73. Kamrin, M.A. (Ed.) 1997. Pesticide profiles: Toxicity, environmental im-pact, and fate. CRC Press, Lewis Publishing, New York.

74. Cooke, A.S. 1981. Tadpoles as indicators of harmful levels of pollution in the field. Environmental Pollution Series A 25:123–133.

75. Raj, T.P., A. Jebanesan, M. Selvanayagam and G.J. Manohar. 1988. Effect of organophosphorus (nuvan) and carbamate (baygon) compounds on *Rana hexadactyla* (Lesson) with a note on body protein and liver glycogen. Geobios 15:25–32.

76. Berrill, M., S. Bertram, A. Wilson, S. Louis, D. Brigham and C. Stromberg. 1993. Lethal and sublethal impacts of pyrethroid insecticides on amphibian em-bryos and tadpoles. Environmental Toxicology and Chemistry 12:525–539.

77. Bauer-Dial, C.A. and N.A. Dial. 1995. Lethal effects of the consumption of field levels of paraquat-contaminated plants on frog tadpoles. Bulletin of Environmental Toxicology 55:870–877.

78. Schuytema, G.S. and A.V. Nebeker. 1998. Comparative toxicity of diuron on survival and growth of Pacific treefrog, bullfrog, red-legged frog, and African clawed frog embryos and tadpoles. Archives of Environmental Contamination and Toxicology 34:370–376.

79. Anderson, R.J. and K.V. Prahlad. 1976. The deleterious effects of fungicides and herbicides on *Xenopus laevis* embryos. Archives of Environmental Contamination and Toxicology 4:312–323.

80. Zavanella, T., N.P. Zaffaroni and E. Arias. 1984. Abnormal limb regenera-tion in adult newts exposed to the fungicide maneb 80. Journal of Toxicology and Environmental Health 13:735–745.

81. Scadding, S.R. 1990. Effects of tributylin oxide on the skeletal structures of developing and regenerating limbs of the axolotl larvae, *Ambystoma mexicanum*. Bulletin of Environmental Contamination and Toxicology 45:574–581.

82. Riley, E.E. and M.R. Weil. 1986. The effects of thiosemicarbazide on de-velopment in the wood frog, *Rana sylvatica*: Concentration effects. Ecotoxicology and Environmental Safety 12:154–160.

Riley, E.E. and M.R. Weil. 1987. The effects of thiosemicarbazide on de-velopment in the wood frog, *Rana sylvatica*: critical exposure length and age sen-sitivity. Ecotoxicology and Environmental Safety 13:202–207.

83. Berrill, M., S. Bertram, L. McGillivray, M. Kolohan and B. Pauli. 1994. Effects of low concentrations of forest-use pesticides on frog embryos and tadpoles. Environmental Toxicology and Chemistry 13:657–664.

84. Linder, G. and B. Grillitsch. 2002. Ecotoxicology of metals. Pp. 325–459. *In* Sparling, D.W., G. Linder and C.A. Bishop (Eds.), Ecotoxicology of Amphibians and Reptiles. Society for Environmental Toxicology and Contaminants (SETAC) Press, Pensacola, Florida.

85. Canton, J.H. and W. Sloof. 1982. Toxicity and accumulation studies of cadmium (Cd^{2+}) with freshwater organisms of different trophic levels. Ecotoxicology and Environmental Safety 6:113–128.

86. Abbasi, S.A. and R. Soni. 1984. Teratogenic effects of chromium (IV) in environment as evidenced by the impact of larvae of amphibian *Rana tigrina*: Implications in the environmental management of chromium. International Journal of Environmental Studies 23:131–137.

87. Perez-Coll, C.S., J. Herkovitz and A. Saliban. 1988. Embryotoxicity of lead to *Bufo arenarum*. Bulletin of Environmental Contamination and Toxicology 41:247–252.

88. Birge, W.J., J.A. Black, A.G. Westerman and B.A. Ramey. 1983. Fish and amphibian embryos: A model system for evaluating teratogenicity. Fundamentals of Applied Toxicology 3:237–242.

89. Sparling, D.W. 2000. Ecotoxicology of organic contaminants to amphibians. Pp. 461–494. *In* Sparling, D.W., G. Linder and C.A. Bishop (Eds.), Ecotoxicology of Amphibians and Reptiles. Society for Environmental Toxicology and Contaminants (SETAC) Press, Pensacola, Florida.

90. Eisler, R. 1987. Polycyclic aromatic hydrocarbon hazards to fish, wildlife, and invertebrates: A synoptic review. U.S. Fish and Wildlife Service, Washington, D.C.

Eisler, R. and A.A. Beslisle. 1996. Planar PCB hazards to fish, wildlife, and invertebrates: a synoptic review. National Biological Service Biological Report Number 31, Washington, D.C.

91. Cooke, A.S. 1970. The effect of p,p'-DDT on tadpoles of the common frog (*Rana temporaria*). Environmental Pollution 1:57–71.

Cooke, A.S. 1972. The effects of DDT, dieldrin, and 2,4-D on amphibian spawn and tadpoles. Environmental Pollution 3:51–68.

Osborn, D., A.S. Cooke and S. Freestone. 1981. History of a teratogenic effect of DDT on *Rana temporaria* tadpoles. Environmental Pollution 25:305–319.

92. McGrath, E.A. and M.M. Alexander. 1979. Observations on the exposure of larval bullfrogs to fuel oil. Transactions of the Northeastern Section of the Wildlife Society 36:45–51.

93. Ouellet, M., J. Bonin, J. Rodrigue, J.-L. DesGranges and S. Lair. 1997. Hindlimb deformities (ectromelia, ectrodactyly) in free-living anurans from agricultural habitats. Journal of Wildlife Diseases 33:95–104.

94. Bly, B.L., M.G. Knutson, M.B. Sandheinrich, B.R. Gray and D.A. Jobe. 2004. Flow cytometry used to assess genetic damage in frogs from farm ponds. Journal of the Iowa Academy of Science 111:67–70.

95. Bryant, S.V., V. French and P.J. Bryant. 1981. Distal regeneration and symmetry. Science 212:993–1002.

96. Sutherland, D. 2005. Parasites of North American frogs. Pp. 109–123. *In* Lannoo, M.J. (Ed.), Amphibian Declines: The Conservation Status of North American Species. University of California Press, Berkeley, California.

97. Johnson, P.T.J. and K.B. Lunde. 2005. Parasite infection and limb malformations: A growing problem in amphibian conservation. Pp. 124–138. *In* Lannoo, M.J. (Ed.), Amphibian Declines: The Conservation Status of United States Species. University of California Press, Berkeley, California.

98. Dave Gardiner, personal communication.

99. Erasmus, D.A. 1972. The Biology of Trematodes. Edward Arnold, London. Haseeb, M.A. and B. Fried. 1997. Modes of transmission of trematode infections and their control. Pp. 31–56. *In* Fried, B. and T.K. Craczyk (Eds.), Advances in Trematode Biology. CRC Press, New York.

But see: Thiemann, G.W. and R.J. Wassersug. 2000. Biased distribution of trematode metacercariae in the nephric system of *Rana* tadpoles. Journal of Zoology, London 252:534–538.

100. Beaver, P.C. 1937. Experimental studies on *Echinostoma revolutum* (Froelich) a fluke from birds and mammals. Illinois Biological Monographs 15:1–96.

101. Taylor, C.N., K.L. Oseen and R.J. Wassersug. 2004. On the behavioral response of *Rana* and *Bufo* tadpoles to echinostomatid cercariae: Implications to synergistic factors influencing trematode infections in anurans. Canadian Journal of Zoology 82:701–706.

See also: Thiemann, G.W. and R.J. Wassersug. 2000. Patterns and consequences of behavioral responses to predators and parasites in *Rana* tadpoles. Biological Journal of the Linnean Society 71:513–528.

102. Sessions, S.K., R.A. Franssen and V.L. Horner. 1999. Morphological clues from multilegged frogs: Are retinoids to blame? Science 284:800–802.

103. Helluy, S. and J.C. Holmes. 1990. Serotonin, octopamine and the clinging behaviour induced by the parasite *Polymorphus paradoxus* (Acancephala) in *Gammarus lacustris* (Crustacea). Canadian Journal of Zoology 68:1214–1220.

Mueller, J.F. 1970. Quantitative relationship between stimulus and response in the growth-promoting effect of *Spirometra mansonoides* Spargana on the hypophysectimized rat. Journal of Parasitology 56:840–842.

104. Converse, K.A., J. Mattsson and L. Eaton-Poole. 2000. Field surveys of Midwestern and Northeastern Fish and Wildlife Service lands for the presence of abnormal frogs and toads. Journal of the Iowa Academy of Science 107: 160–167.

See also: Schoff, P.K., C.M. Johnson, A.M. Schotthoefer, J.E. Murphy, C. Lieske, R.A. Cole, L.B. Johnson and V.R. Beasley. 2003. Prevalence of skeletal and eye malformations in frogs from north-central United States: estimations based on collections from randomly selected sites. Journal of Wildlife Diseases 39:510–521.

105. Skelly, D.K., S.R. Bolden, L.K. Freidenburg, N.A. Freidenfelds and R. Levey. 2007. *Ribeiroia* infection is not responsible for Vermont amphibian deformities. EcoHealth 4:156-163.

106. Johnson, P.T.J. and D.R. Sutherland. 2003. Amphibian deformities and *Ribeiroia* infection: an emerging helminthiasis. Trends in Parasitology 19:332–335.

107. Ecoregions indicated are U.S. E.P.A. Level III ecoregions as defined by the following article: United States Environmental Protection Agency (USEPA). 2000. Level III ecoregions of the continental United States (revision of Omernik, 1987), Map M-1. U.S.E.P.A. National Health and Environmental Effects Research Laboratory, Corvallis, Oregon.

108. Holland, M.P., D.K. Skelly, M. Kashgarian, S.R. Bolden, L.M. Harrison and M. Capello. 2007. Echinostome infection in green frogs is stage and age dependent. Journal of Zoology 271:455–462.

Schotthoefer, A.M., R.A. Cole and V.R. Beasley. 2002. Relationship of tadpole stage to location of echinostome cercariae encystment and the consequences for tadpole survival. Journal of Parasitology 89:475–482.

109. Dare, O.K., P.L. Rutherford and M.R. Forbes. 2006. Rearing density and susceptibility of *Rana pipiens* metamorphs to cercariae of a digenetic trematode. Journal of Parasitology 92:543–547.

110. Fried, B., P.L. Pane and A. Reddy. 1997. Experimental infection of *Rana pipiens* tadpoles with *Echinostoma trivolvis* cercariae. Parasitology Research 83:666–669.

111. Kiesecker, J.M. and D.K. Skelly. 2001. Effects of disease and pond drying on gray tree frog growth, development, and survival. Ecology 82:1956–1963.

CHAPTER 6: HUMAN MALFORMATIONS AND CAUSES

1. Frost, R. 1920. The Oven Bird. *In* Mountain Interval. Henry Holt, New York.

2. Smith, D.W. 1976. Recognizable Patterns of Human Malformation: Genetic, Embryologic and Clinical Aspects. Second Edition. Volume VII in the Series, Major Problems in Clinical Pediatrics, Schaffer, A.J., Consulting Editor. W.B. Saunders Company, Philadelphia, Pennsylvania.

3. National Center for Health Statistics, Hyattsville, Maryland, a component of the U.S. Department of Health and Human Services' Centers for Disease Control and Prevention, Atlanta, Georgia.

4. Rates of chromosomal anomalies vary by source and should be considered estimates, not hard facts.

5. Bingley, P.J., I.F. Douek, C.A. Rogers and E.A.M. Gale. 2000. Influence of maternal age at delivery and birth order on risk of type 1 diabetes in childhood: Prospective population based family study. British Medical Journal 321:420–424.

Croen, L.A. and G.M. Shaw. 1995. Young maternal age and congenital malformations: A population-based study. American Journal of Public Health 85:710–713.

Ferguson-Smith, M.A. and R.W. Yates. 1984. Maternal age specific rates for chromosomal aberrations and factors influencing them: Report of a collaborative European study in 52,965 amniocenteses. Prenatal Diagnosis 4:5–44.

Fisch, H., R.J. Golden, G.L. Libersen, G.S. Hyun, P. Madsen, M.I. New and T.W. Hensle. 2001. Maternal age as a risk factor for hypospadias. Journal of Urology 165:934–936.

Hansen, J.P. 1986. Older maternal age and pregnancy outcome: A review of the literature. Obstetrical and Gynecological Survey 41:726–742.

Hay, S. and H. Barbano. 1972. Independent effects of maternal age and birth order on the incidence of selected congenital malformations. Teratology 6:271–279.

6. Bound, J.P., P.W. Harvey and B.J. Francis. 1989. Seasonal prevalence of major congenital malformations in the Fylde of Lancashire 1957–1981. Journal of Epidemiology and Community Health 43:330–342.

Coupland, M.A. and A.I. Coupland. 1988. Seasonality, incidence, and sex distribution of cleft lip and palate births in Trent Region, 1973–1982. Cleft Palate Journal 25:33–37.

Davies, B.R. 2000. The seasonal conception of lethal congenital malformations. Archives of Medical Research 31:589–591.

Garry, V.F., M.E. Harkins, L.L. Erickson, L.K. Long-Simpson, S.E. Holland and B.L. Burroughs. 2002. Birth defects, season of conception, and sex of children born to pesticide applicators living in the Red River Valley of Minnesota, USA. Environmental Health Perspectives 110:441–449.

7. North, K. and J. Golding. 2000. A maternal vegetarian diet in pregnancy is associated with hypospadias. BJU [British Journal of Urology] International 85: 107–113.

8. Rosa, F.W., A.L. Wilk and F.O. Kelsey. 1986. Teratogen update: Vitamin A congeners. Teratology 33:355–364.

9. Aberg, A., L. Westbom and B. Kallen. 2001. Congenital malformations among infants whose mothers had gestational diabetes or preexisting diabetes. Early Human Development 61:85–95.

Freij, B.J., M.A. South and J.L. Sever. 1988. Maternal Rubella and the congenital Rubella syndrome. Clinical Perinatology 15:247–257.

10. Burton, B.K., C.J. Schulz and L.I. Burd. 1992. Limb abnormalities associated with chorionic villus sampling (CVS). Pediatric Research 31:69A.

Hsieh, F.J., M.K. Shyu, B.C. Sheu, S.P. Lin, C.P. Chen and F.Y. Huang. 1995. Limb defects after chorionic villus sampling. Obstetrics and Gynecology 85:84–88.

11. Silver, R.I., R. Rodriguez, T.S. Chang and J.P. Gearhart. 1999. In vitro fertilization is associated with an increased risk of hypospadias. Journal of Urology161:1954–1957.

12. Arpino, C., S. Brescianini, E. Robert, E.E. Castilla, G. Cocchi, M.C. Cornel, C. de Vigan, P.A. Lancaster, P. Merlob, Y. Sumiyoshi, G. Zampino, C. Renzi, A. Rosano and P. Mastroiacovo. 2000. Teratogenic effects of antiepileptic drugs: Use of International Database on Malformations and Drug Exposure. Epilepsia 41:1436–1443.

Castilla, E.E., P. Ashton-Prolla, E. Barreda-Mejia, D. Brunoni, D.P. Cavalcanti, J. Correa-Neto, J.L. Delgadillo, M.G. Dutra, T. Felix, A. Giraldo, N. Juarez, J.S. Lopez-Camelo, J. Nazer, I.M. Orioli, J.E. Paz, M.A. Pessoto, J.M. Pina-Neto, R. Quadrelli, M. Rittler, S. Rueda, M. Saltos, O. Sanchez and L. Schuler. 1996. Thalidomide, a current teratogen in South America. Teratology 54:273–277.

Czeizel, A.E., M. Rockenbauer, H.T. Sorensen and J. Olsen. 2001. A population-based case-control teratologic study of ampicillin treatment during pregnancy. American Journal of Obstetrics and Gynecology 185:140–147.

Czeizel, A.E. and M. Rockenbauer. 2000. A population-based case-control teratologic study of oral oxytetracycline treatment during pregnancy. European Journal of Obstetrics and Gynecology and Reproductive Biology 88:27–33.

Correy, J.F., N.M. Newman, J.A. Collins, E.A. Burrows, R.F. Burrows and J.T. Curran. 1991. Use of prescription drugs in the first trimester and congenital malformations. Australian and New Zealand Journal of Obstetrics and Gynaecology 31:340–344.

Ericson, A. and B.A. Kallen. 2001. Nonsteroidal anti-inflammatory drugs in early pregnancy. Reproductive Toxicology 15:371–375.

Golding, J., S. Vivian and J.A. Baldwin. 1983. Maternal anti-nauseants and clefts of lip and palate. Human Toxicology 2:63–73.

Hernandez-Diaz, S., M.M. Werler, A.M. Walker and A.A. Mitchell. 2000. Folic acid antagonists during pregnancy and the risk of birth defects. New England Journal of Medicine 343:1608–1614.

Kaneko, S. and T. Kondo. 1995. Antiepileptic agents and birth defects. CNS [Central Nervous System] Drugs 3:41–55.

Klip, H., J. Verloop, J.D. van Gool, M.E.T.A. Koster, C.W. Burger and F.E. van Leeuwen. 2002. Hypospadias in sons of women exposed to diethyl-stilbestrol (DES) in utero: A cohort study. Lancet 359:1102–1107.

Kricker, A., J.W. Elliott, J.M. Forrest and J. McCredie. 1986. Congenital limb reduction deformities and use of oral contraceptives. American Journal of Obstetrics and Gynecology 155:1072–1078.

Lindhout, D. and J.G. Omtzigt. 1994. Teratogenic effects of antiepileptic drugs: Implications for the management of epilepsy in women of childbearing age. Epilepsia 35, Supplement 4:S19–28.

Matalon, S., S. Schechtman, G. Goldzweig and A. Ornoy. 2002. The teratogenic effect of carbamzepine: A meta-analysis of 1255 exposures. Reproductive Toxicology 16:9–17.

Milkovich, L. and B.J. van der Berg. 1977. Effects of antenatal exposure to anorectic drugs. American Journal of Obstetrics and Gynecology 129:637–642.

Okada, T., T. Tomoda, H. Hisakawa and T. Kurashige. 1995. Fetal valproate syndrome with reduction deformity of limb. Acta Paediatrica Japonica 1995 37:58–60.

Orioli, I.M. and E.E. Castilla. 2000. New associations between prenatal exposures to drugs and malformations. American Journal of Human Genetics 67(4 Supplement 2):175.

Rubin, J.D., C. Loffredo, A. Correa-Villaseñor, C. Ferencz and the Baltimore-Washington Infant Study Group. 1991. Prenatal drug use and congenital cardiovascular malformations. Teratology 43:423.

Saxen, I. 1975. Associations between oral clefts and drugs taken during pregnancy. International Journal of Epidemiology 4:37–44.

Schatz, M. 2001. The efficacy and safety of asthma medications during pregnancy. Seminars in Perinatology 25:145–152.

Sharony, R., A. Garber, D. Viskochil, R. Schreck, L.D. Platt, R. Ward, B.A. Buehler and J.M. Graham, Jr. 1993. Preaxial ray reduction defects as part of valproic acid embryofetopathy. Prenatal Diagnosis 13:909–918.

Torfs, C.P., E.A. Katz, T.F. Bateson, P.K. Lam and C.J.R. Curry. 1996. Maternal medications and environmental exposures as risk factors for gastroschisis. Teratology 54:84–92.

Werler, M.M., A.A. Mitchell and S. Shapiro. 1992. First trimester maternal medication use in relation to gastroschisis. Teratology 45:361–367.

13. Cordero, J.F., G.P. Oakley, F. Greenberg and L.M. James. 1981. Is bendectin a teratogen? Journal of the American Medical Association 245:2307–2310.

Ericson, A. and B.A. Kallen. 2001. Nonsteroidal anti-inflammatory drugs in early pregnancy. Reproductive Toxicology 15:371–375.

Hendrickx, A.G., M. Cukierski, S. Prahalada, G. Janos and J. Rowland. 1985. Evaluation of bendectin embryotoxicity in nonhuman primates: I. Ventricular septal defects in prenatal macaques and baboon. Teratology 32:179–189.

Kozer, E., A. Costei, R. Boskovic, I. Nulman, S. Nikfar and G. Koren. 2002. Association of aspirin consumption during the first trimester of pregnancy with congenital anomalies: a meta-analysis. Pediatric Research 51(4 Part 2):68A–69A.

Orioli, I.M. and E.E. Castilla. 2000. New associations between prenatal exposures to drugs and malformations. American Journal of Human Genetics 67(4 Supplement 2):175.

Torfs, C.P., E.A. Katz, T.F. Bateson, P.K. Lam and C.J.R. Curry. 1996. Maternal medications and environmental exposures as risk factors for gastroschisis. Teratology 54:84–92.

Werler, M.M., A.A. Mitchell and S. Shapiro. 1992. First trimester maternal medication use in relation to gastroschisis. Teratology 45:361–367.

Zierler, S. and K.J. Rothman. 1985. Congenital heart disease in relation to maternal use of bendectin and other drugs in early pregnancy. New England Journal of Medicine 313:347–352.

14. Battin, M., S. Albersheim and D. Newman. 1995. Congenital genitourinary tract abnormalities following cocaine exposure in utero. American Journal of Perinatology 12:425–428.

Hoyme, H.E., K.L. Jones and S.D. Dixon, T. Jewett, J.W. Hanson, L.K. Robinson, M.E. Msall and J.E. Allanson. 1990. Prenatal cocaine exposure and fetal vascular disruption. Pediatrics 85:743–747.

15. Carr, B.K. 1997. Congenital limb reduction defects in infants: A look at possible associations with maternal smoking and hypertension. National Technical Information Service Technical Report (NTIS/AD-A330 296), Springfield, Virginia, 29 pp.

Chung, K.C., C.P. Kowalski, H.M. Kim, M. Hyungjin, and S.R. Buchman. 2000. Maternal cigarette smoking during pregnancy and the risk of having a child with cleft lip/palate. Plastic Reconstructive Surgery105:485–491.

Eskenazi, B. and R. Castorina. 1999. Association of prenatal maternal or postnatal child environmental tobacco smoke exposure and neurodevelopmental and behavioral problems in children. Environmental Health Perspectives 107:991–1000.

Lorente, C., S. Cordier and J. Goujard, S. Aymé, F. Bianchi, E. Calzolari, H.E.K. De Walle, R. Knill-Jones and the Occupational Exposure and Congenital Malformations Working Group. 2000. Tobacco and alcohol use during pregnancy and risk of oral clefts. Occupation Exposure and Congenital Malformation Working Group. American Journal of Public Health 90:415–419.

Singh, S., E. Barrett, R. Kalra, S. Razani-Boroujerdi, R.J. Langley, V. Kurup, Y. Tesfaigzi and M. Sopori. 2003. Prenatal cigarette smoke decreases lung cAMP and increases airway hyperresponsiveness. American Journal of Respiratory and Critical Care Medicine 168:342–347.

Wang, X., B. Zuckerman, C. Pearson, G. Kaufman, C. Chen, G. Wang, T. Niu, P.H. Wise, H. Bauchner and X. Xu. 2002. Maternal cigarette smoking, metabolic gene polymorphism, and infant birth weight. Journal of the American Medical Association 287:195–202.

16. Lorente, C., S. Cordier and J. Goujard, S. Aymé, F. Bianchi, E. Calzolari, H.E.K. De Walle, R. Knill-Jones and the Occupational Exposure and Congenital Malformations Working Group. 2000. Tobacco and alcohol use during pregnancy and risk of oral clefts. Occupation Exposure and Congenital Malformation Working Group. American Journal of Public Health 90:415–419.

Streissguth, A.P., J.M. Aase, S.K. Clarren, S.P. Randels, R.A. LaDue and D.F. Smith. 1991. Fetal alcohol syndrome in adolescents and adults. Journal of the American Medical Association 265:1961–1967.

17. Bao, Y.S., S. Cai, S.F. Zhao, X.C. Xhang, M.Y. Huang, O. Zheng and Jiang. 1991. Birth defects in the offspring of female workers occupationally exposed to carbon disulfide in China. Teratology 43:451–452.

Barlow, S.M. and F.M. Sullivan. 1982. Reproductive hazards associated with different occupational groups. Pp. 32–39. In Barlow, S.M. and F.M. Sullivan (Eds.), Reproductive Hazards of Industrial Chemicals: An Evaluation of Animal and Human Data. Academic Press, New York.

Bianchi, F., D. Cianciulli, A. Pierini and S. Costantini. 1997. Congenital malformations and maternal occupation: A registry based case-control study. Occupational and Environmental Medicine 54:223–228.

Blatter, B. and N. Roeleveld. 1996. Spina bifida and parental occupation in a Swedish register-based study. Scandinavian Journal of Work, Environment and Health 22:433–437.

Brender, J.D. and L. Suarez. 1990. Paternal occupation and anencephaly. American Journal of Epidemiology131:517–521.

Brender, J., L. Suarez, K. Hendricks, R.A. Baetz and R. Larsen. 2001. Parental occupation and neural tube defect-affected pregnancies among Mexican Americans. Journal of Occupational and Environmental Medicine 44:650–656.

Cordier, S., A. Bergeret, J. Goujard, M.C. Ha, S. Ayme, F. Bianchi, E. Calzolari, H.E.K. De Walle, R. Knill-Jones, S. Candela, I. Dale, B. Dananche, C. De Vignan, J. Fevotte, G. Kiel and L. Mandereau. 1997. Congenital malformation and maternal occupational exposure to glycol ethers. Epidemiology 8:355–363.

Cordier, S., M.C. Ha, S. Ayme and J. Goujard. 1992. Maternal occupational exposure and congenital malformations. Scandinavian Journal of Work, Environment and Health 18:11–17.

Correa-Villaseñor, A., P.D. Wilson, C. Loffredo, C. Ferencz and the Baltimore-Washington Infant Study Group. 1991. Cardiovascular malformation and prenatal environmental exposures. Pediatric Research 29 (4 Part 2):17A.

Dimich-Ward, H., C. Hertzman, K. Teschke, R. Hershler, S.A. Marion, A. Ostry and S. Kelly. 1996. Reproductive effects of paternal exposure to chlorophenate wood preservatives in the sawmill industry. Scandinavian Journal of Work, Environment and Health 22:267–273.

Engel, L.S., E.S. O'Meara and S.M. Schwartz. 2000. Maternal occupation in agriculture and risk of limb defects in Washington State, 1980–1993. Scandinavian Journal of Work, Environment and Health 26:193–198.

García, A.M. and T. Fletcher. 1998. Maternal occupation in the leather industry and selected congenital malformations. Occupation and Environmental Medicine 55:284–286.

García, A.M., T. Fletcher, F.G. Benavides and E. Orts. 1999. Parental agricultural work and selected congenital malformations. American Journal of Epidemiology 149:64–74.

Irgens, A., K. Krüger, A.H. Skorve and L.M. Irgens. 1998. Reproductive outcome in offspring of parents occupationally exposed to lead in Norway. American Journal of Industrial Medicine 34:431–437.

Irgens, A., K. Krüger, A.H. Skorve and L. Irgens. 2000. Birth defects and paternal occupational exposure: Hypotheses tested in a record linkage based dataset. Acta Obstetricia et Gynecologica Scandinavica 79:465–470.

Kristensen, P., L.M. Irgens, A. Andersen, A.S. Bye and L. Sundheim. 1997. Birth defects among offspring of Norwegian farmers, 1967–1991. Epidemiology 8:537–544.

Matte, T.D., J. Mulinare and J.D. Erickson. 1993. Case-control study of congenital defects and parental employment in health care. American Journal of Industrial Medicine 24:11–23.

Olshan, A.F., K. Teschke and P.A. Baird. 1991. Paternal occupation and congenital anomalies in offspring. American Journal of Industrial Medicine 20:447–475.

Schwartz, D.A. and J.P. LoGerfo. 1988. Congenital limb reduction defects in the agricultural setting. American Journal of Public Health 78:654–658.

Sever, L.E., T.E. Arbuckle and A. Sweeney. 1997. Reproductive and developmental effects of occupational pesticide exposure: the epidemiologic evidence. Occupational Medicine: State of the Art Reviews 12:305–325.

Shaw, G.M., C.R. Wasserman, C.D. O'Malley, V. Nelson and R.J. Jackson. 1999. Maternal pesticide exposure from multiple sources and selected congenital anomalies. Epidemiology 10:60–66.

18. Bobak, M. 2000. Outdoor air pollution, low birth weight, and prematurity. Environmental Health Perspectives 108:173–176.

Dejmek, J., S.G. Selevan, I. Benes, I. Solansky and R.J. Sram. 1999. Fetal growth and maternal exposure to particulate matter during pregnancy. Environmental Health Perspectives 107:475–480.

Ha, E.H., Y.C. Hong, B.E. Lee, B.H. Woo, J. Schwartz and D.C. Christiani. 2001. Is air pollution a risk factor for low birth weight in Seoul? Epidemiology 12:643–648.

Maisonet, M., T. Bush, A. Correa and J. Jaakkola. 2001. Relation between ambient air pollution and low birth weight in the northeastern United States. Environmental Health Perspectives 109 (Supplement 3):351–356.

Ritz, B., F. Yu, G. Chapa and S. Fruin. 2000. Effect of air pollution on preterm birth among children born in Southern California between 1989 and 1993. Epidemiology 11:502–511.

Wang, X., H. Ding, L. Ryan and X. Xu. 1997. Association between air pollution and low birth weight: a community-based study. Environmental Health Perspectives 105:514–520.

Xu, X., H. Ding and X. Wang. 1995. Acute effects of total suspended particles and sulfur dioxides on preterm delivery: a community based cohort study. Archives of Environmental Health 50:407–415.

19. Bove, F.J., M.C. Fulcomer, J.B. Klotz, J. Esmart, E.M. Dufficy and J.E. Savrin. 1995. Public drinking water contamination and birth outcomes. American Journal of Epidemiology 141:850–862.

Cedergren, M.I., A.J. Selbing, O. Lofman and B.A.J. Kallen. 2002. Chlorination byproducts and nitrate in drinking water and risk for congenital cardiac defects. Environmental Research 89:124–130.

Engel, R.R. and A.H. Smith. 1994. Arsenic in drinking water and mortality from vascular disease: An ecologic analysis in 30 countries in the United States. Archives of Environmental Health 49:418–427.

Hwang, B.F., P. Magnus and J.J. Jaakkola. 2002. Risk of specific birth defects in relation to chlorination and the amount of natural organic matter in the water supply. American Journal of Epidemiology 156:374–382.

Lonky, E., J. Reihman, T. Darvill, J. Mather and H. Daly. 1996. Neonatal behavioral assessment scale performance in humans influenced by maternal consumption of environmentally contaminated Lake Ontario fish. Journal of Great Lakes Research 22:198–212.

Magnus, P., J. Jaakkola, A. Skrondal, J. Alexander, G. Bechger, T. Krogh and E. Dybing. 1999. Water chlorination and birth defects: A nation-wide registry based study. Epidemiology 10:513–517.

Munger, R., P. Isacson, M. Kramer, J. Hanson, T. Burns, K. Cherryholmes and W. Hausler Jr. 1992. Birth defects and pesticide-contaminated water supplies in Iowa. American Journal of Epidemiology 136:959.

Zierler, S., M. Theodore, A. Cohen and K.J. Rothman.1988. Chemical quality of maternal drinking water and congenital heart disease. International Journal of Epidemiology 17:589–594.

20. Andrews, J.S. Jr. 1992. Polychlorodibenzodioxins and Polychlorodibenzofurans. Pp. 756–761. In Sullivan, J.B., Jr. and G.R. Krieger (Eds.), Hazardous Materials Toxicology, Clinical Principals of Environmental Health. Williams and Wilkins, Baltimore, Maryland.

Fara, G.M. and G. Del Corno. 1985. Pregnancy outcome in the Seveso area after TCDD contamination. Progress in Clinical and Biological Research 163B:279–285.

Landrigan, P.J., C.B. Schechter, J.M. Lipton, M.C. Fahs and J. Schwartz. 2002. Environmental pollutants and disease in American children: Estimates of morbidity, mortality, and costs for lead poisoning, asthma, cancer, and developmental disabilities. Environmental Health Perspectives 110:721–728.

Perera, F., V. Rauh, T. We-Yann, P. Kinney, D. Camann, D. Barr, T. Bernert, R. Garkinkel, Y-H. Tu, D. Diaz, J. Dietrich and R.K. Whyatt. 2003. Effects of transplacental exposure to environmental pollutants on birth outcomes in a multiethnic population. Environmental Health Perspectives 111:201–205.

Shaw, G.M., J. Schulman, J.D. Frisch, S.K. Cummins and J.A. Harris. 1992. Congenital malformations and birthweight in areas with potential environmental contamination. Archives of Environmental Health 47:147–154.

Sweeny, A. 1994. Reproductive epidemiology of dioxins. Pp. 549–583. In Schecter, A. (Ed.), Dioxins and Health. Plenum Press, New York.

Stewart, P.W., J. Reihman, E.I. Lonky, T.J. Darvill and J. Pagano. 2003. Cognitive development in preschool children prenatally exposed to PCBs and MeHg. Neurotoxicology and Teratology 25:11–22.

Vos, J., C. de Heer and H. van Lovern. 1997/98. Immunotoxic effects of TCDD and toxic equivalency factors. Teratogenesis, Carcinogenesis, and Mutagenesis 17:275–284.

Weisglas-Kuperus, N., S. Patandin, G. Berbers, T.C.J. Sas, P.G.H. Mulder, P.J.J. Sauer and H. Hooijkas. 2000. Immunologic effects of background exposure to polychlorinated biphenyls and dioxins in Dutch preschool children. Environmental Health Perspectives 108:1203–1207.

21. Bell, E., I. Hertz-Picciotto and J. Beaumont. 2001. A case-control study of pesticides and fetal death due to congenital anomalies. Epidemiology 12:148–156.

Committee to Review the Health Effects in Vietnam Veterans of Exposure to Herbicides, Institute of Medicine. Conclusions about health outcomes: health outcomes with limited/suggestive evidence of an association. 1996. Veterans and Agent Orange: Update 1996 pages 1–7 to 1–9, National Academy Press, Washington, D.C.

Guillette, E.A., M.M. Meza, M.G. Aguilar, A.D. Soto and I.E. Garcia. 1998. An anthropological approach to the evaluation of preschool children exposed to pesticides in Mexico. Environmental Health Perspectives 106:347–353.

Hayes, W.J., Jr. 1982. Pesticides Studied in Man. Williams and Wilkins, London, England.

Longnecker, M.P., M.A. Klebanoff, H. Zhou and J.W. Brock, 2001. Association between maternal serum concentration of the DDT metabolite DDE and preterm and small-for-gestational-age babies at birth. Lancet 358:110–114.

Munger, R., P. Isacson, M. Kramer, J. Hanson, T. Burns, K. Cherryholmes and W. Hausler Jr. 1992. Birth defects and pesticide-contaminated water supplies in Iowa. American Journal of Epidemiology136:959.

Shaw, G.M., C.R. Wasserman, C.D. O'Malley, V. Nelson and R.J. Jackson. 1999. Maternal pesticide exposure from multiple sources and selected congenital anomalies. Epidemiology 10:60–66.

22. Bellinger, D., A. Leviton, E. Allred and M. Rabinowitz. 1994. Pre- and postnatal lead exposure and behavior problems in school-age children. Environmental Research 66:12–30.

Correa, A., C. Loffredo, C. Ferencz and P.D. Wilson. 1990. Lead and solvent exposure during pregnancy: possible risk of cardiovascular malformations (CVM). Teratology 41:545.

Crinella, F., E. Cordova and J. Ericson. 1998. Manganese, aggression, and attention-deficit hyperactivity disorder. Neurotoxicology 19:468–469.

Engel, R.R. and A.H. Smith. 1994. Arsenic in drinking water and mortality from vascular disease: an ecologic analysis in 30 countries in the United States. Archives of Environmental Health 49:418–427.

Landrigan, P.J., C.B. Schechter, J.M. Lipton, M.C. Fahs and J. Schwartz. 2002. Environmental pollutants and disease in American children: estimates of morbidity, mortality, and costs for lead poisoning, asthma, cancer, and developmental disabilities. Environmental Health Perspectives 110:721–728.

Vinceti, M., S. Rovesti and M. Bergomi. 2001. Risk of birth defects in a population exposed to environmental lead pollution. Science of the Total Environment 278:23–30.

23. Croen, L.A., G.M. Shaw, L. Sanbonmatsu, S. Selvin and P.A. Buffler. 1997. Maternal residential proximity to hazardous waste sites and risk for selected congenital malformations. Epidemiology 8:347–354.

Dolk, H., M. Vrijheid, B. Armstrong, L. Abramsky, F. Bianchi, E. Garne, V. Nelen, E. Robert, J.E. Scott, D. Stone and R. Tenconi. 1998. Risk of congenital anomalies near hazardous-waste landfill sites in Europe: The EURO-HAZCON study. Lancet 352:423–427.

Dolk, H. and M. Vrijheid. 2003. The impact of environmental pollution on congenital anomalies. British Medical Bulletin 68:25–45.

Orr, M.F. 1999. Birth defects among children of racial or ethnic minority born to women living in close proximity to hazardous waste sites; California, 1983–1988. National Technical Information Service Technical Report (NTIS/PB99-139990), Springfield, Virginia, 76 pp.

24. Greenland, S. and D.L. Ackerman. 1995. Clomiphene citrate and neural tube defects: A pooled analysis of controlled epidemiologic studies and recommendations for future studies. Fertility and Sterility 64:936–941.

Heinonen, O.P., D. Slone, R.R. Monson, E.B. Hook and S. Shapiro. 1977. Cardiovascular birth defects and antenatal exposure to female sex hormones. New England Journal of Medicine 296:67–70.

Hook, E.B. 1994. Cardiovascular birth defects and prenatal exposure to female sex hormones: A reevaluation of data reanalysis from a large prospective study. Teratology 49:162–166.

Hwang, S-J., T.H. Beaty, S.R. Panny, N.A. Street, J.M. Joseph, S. Gordon, I. McIntosh and C.A. Francomano. 1995. Association study of transforming growth factor alpha (TGFa) TaqI polymorphism and oral clefts: Indication of a gene-environment interaction in a population-based sample of infants with birth defects. American Journal of Epidemiology 141:629–636.

Mori, C. 2001. Possible effects of endocrine disruptors on the reproductive system. Teratology 2001 63:9A.

National Research Council. 1999. Hormonally active agents in the environment. National Academy Press, Washington D.C.

Park-Wyllie, L., P. Mazzotta, A. Pastuszak, M.E. Moretti, L. Beique, L. Hunnisett, M.H. Friesen, S. Jacobson, S. Kasapinovic, D. Chang, O. Diav-Citrin, D. Chitayat, I. Nulman, T.R. Einarson and G. Koren. 2000. Birth defects after maternal exposure to corticosteroids: Prospective cohort study and meta-analysis of epidemiological studies. Teratology 62:385–392.

Rothman, K.J., D.C. Fyler, A. Goldblatt and M.R. Kreidberg. 1979. Exogenous hormones and other drug exposures of children with congenital heart disease. American Journal of Epidemiology 109:433–439.

Toppari, J. 2002. Environmental endocrine disrupters and disorders of sexual differentiation. Seminars in Reproductive Medicine 20:305–312.

25. Donald, J., K. Hooper and C. Hopenheyn-Rich. 1991. Developmental toxicity of toluene: Evidence from animal and human studies. Archives of Environmental Health 46:125.

26. Wilcox, A.J., C.R. Weinberg, J.F. O'Connor, D.D. Baird, J.P. Schlatterer, R.E. Canfield, E.G. Armstrong and B.C. Nisula. 1988. Incidence of early loss pregnancy. New England Journal of Medicine 319:189–194.

27. See for example: http://www.amnioticbandsyndrome.com/

28. See for example, the following paper: Cohen, M.M., Jr. 2001. Frog decline, frog malformations, and a comparison of frog and human health. American Journal of Medical Genetics 104:101–109.

See also: van der Shalie, W.H., H.S. Gardner Jr., J.A. Bantle, C.T. De Rosa, R.A. Finch, J.S. Reif, R.H. Reuter, L.C. Backer, J. Burger, L.C. Folmar and W.S. Stokes. 1999. Animals as sentinels of human health hazards of environmental chemicals. Environmental Health Perspectives 107:309–315.

29. Lowcock, L.A., T.F. Sharbel, J. Bonin, M. Ouellet, J. Rodrigue and J.-L. DesGranges. 1997. Flow cytometric assay for in vivo genotoxic effects of pesticides in green frogs (Rana clamitans). Aquatic Toxicology 38:241–255.

30. Among North American frog species, the similarly-appearing Hyla chrysoscelis is diploid, Hyla versicolor is tetraploid. Among salamanders, a number of forms that we call "unisexual hybrids" related to Ambystoma jeffersonianum or A. laterale are distinguished by different chromosomal numbers and species contributions (see species accounts in Lannoo, M.J. [Ed.]. 2005. Amphibian Declines: The Conservation Status of United States Species. University of California Press, Berkeley, California).

31. See for example: Mizgireuv, I.V., N.L. Flax, L.J. Borkin and V.V. Khudoley. 1984. Dysplastic lesions and abnormalities in amphibians associated with environmental conditions. Neoplasma 31:175–181.

See also: Hopkins, W.A., J. Congdon and J.K. Ray. 2000. Incidence and impact of axial malformations in larval bullfrogs (Rana catesbeiana) developing in sites polluted by a coal-burning power plant. Environmental Toxicology and Chemistry 19:862–868.

Rowe, C.L., O.M. Kinney and J.D. Congdon. 1998. Oral deformities in tadpoles of the bullfrog (Rana catesbeiana) caused by conditions in a polluted habitat. Copeia 1998:244–246.

Rowe, C.L., O.M. Kinney, A.P. Fiori and J.D. Congdon. 1996. Oral deformities in tadpoles (Rana catesbeiana) associated with coal ash deposition: Effects on grazing ability and growth. Freshwater Biology 36:723–730.

32. Kolpin, D.W., E.T. Furlong, M.T. Meyer, E.M. Thurman, S.D. Zuagg, L.B. Barber and H.T. Buxton. 2002. Pharmaceuticals, hormones, and other organic wastewater contaminants in U.S. streams, 1999–2000: A national reconnaissance. Environmental Science and Technology 36:1202–1211.

33. In addition to the citations in Chapter 5, see: Sone, K., M. Hinago, A. Kitayama, J. Morokuma, N. Ueno, H. Watanabe and T. Iguchi. 2004. Effects of 17b-estradiol, nonylphenol, and bisphenol-A on developing *Xenopus laevis* embryos. General and Comparative Endocrinology 138:228–236.

See also: Sower, S.A., K.L. Reed and K.J. Babbitt. 2000. Limb malformations and abnormal sex hormone concentrations in frogs. Environmental Health Perspectives 108:1085–1090.

CHAPTER 7: SOLUTIONS

1. Reprinted in Feynman, R.P. 1988. What Do You Care What Other People Think? W.W. Norton, New York, p. 248.

2. Lowcock, L.A., T.F. Sharbel, J. Bonin, M. Ouellet, J. Rodrigue and J.-L. DesGranges. 1997. Flow cytometric assay for in vivo genotoxic effects of pesticides in green frogs (*Rana clamitans*). Aquatic Toxicology 38:241–255.

3. Brown, M.B. 1979. Love Canal and what it says about the poisoning of America. The Atlantic. December.

4. Albert, R.E. 1980. Panel review of Biogenics Corporation study of chromosome abnormalities in Love Canal residents. Report by Environmental Protection Agency panel convened on 27 May, 1980. Panel members were A.D. Auerbach, M. Cohen, K. Hirschhorn, H. Klinger, P. Nowell and N. Wald. Observers were S. Green and P. Voytek. A copy of this report was given to M.J.L. by R. McKinnell.

Kolata, G.B. 1980. Love Canal: False alarm caused by botched study. Science 208:1239–1242.

5. Kolata, G.B. 1980. Love Canal: False alarm caused by botched study. Science 208:1239–1242.

Picciano, D. 1980. Letters: Love Canal chromosomal study. Science 209: 754, 756.

Shaw, M.W. 1980. Letters: Love Canal chromosomal study. Science 209:751–752.

6. Letter from Dr. D. Jack Kilian to Dr. Vilma Hunt, Deputy Assistant Administrator for Health and Ecological Effects, US EPA dated 5 June 1980. A copy of this letter was given to MJL by R. McKinnell.

7. Picciano, D. 1980. Letters: Love Canal chromosomal study. Science 209:754, 756.

8. Shaw, M.W. 1980. Letters: Love Canal chromosomal study. Science 209:751–752.

9. Gardiner, D.M. and D.M. Hoppe. 1999. Environmentally induced limb malformations in mink frogs (*Rana septentrionalis*). Journal of Experimental Zoology 284:207–216, p. 211.

10. The response of organisms to stress is variable and is generally considered to be negative. There are exceptions. Veterans often claim they never felt more alive then when they were in combat and facing death. In vertebrate adults, there is a well understood stress response involving what is called the hypothalamus-pituitary-adrenal axis (the hypothalamus is an old brain region that controls drives such as hunger, thirst, and procreation; the pituitary is the "master gland" controlling hormones [hormones do their work through the endocrine system]; the paired adrenal glands sit on top of the kidneys). The hypothalamus triggers the pituitary, which releases corticotropin releasing hormone (CRH). In the endocrine system, CRH acts on the adrenal glands, which secrete adrenocorticotropin (ACTH) and a class of chemicals called catecholamines. This pathway has generated interest following the demonstration that stress can influence disease outcomes (Fulford and Jessop, p. 176; see reference at the end of this endnote). In fact, immune system cells are believed to produce the structural equivalent of CRH found in neuroendocrine cells (Fulford and Jessop, p. 175; below).

In 1936, Seyle developed his "doctrine of non-specificity," which defines stress as the nonspecific response of the body to any demand. This concept became the unitary stress syndrome and deeply influenced views in the developing field of neuroendocrinology (Goldstein and colleagues, p. 249; see below). But this is science, and we know what happens when new data are collected and science speaks truth to doctrine. The unitary stress syndrome was challenged in 1998 when Pacak and colleagues demonstrated that different stressors can produce diverse neurochemical and neuroendocrine effects. This new view of Pacak and colleagues extends this finding and proposes that each stressor has a unique neurochemical signal with a distinct response, and seems to be displacing the unitary stress syndrome (Palkovits, p.1; see below).

Stress in embryos, especially amphibian embryos, which do not develop in a uterus under the influence of a mother's endocrine and immune systems, is harder to determine. Presumably, young embryos have no ability to mount either an endocrine or an immune response to stress, but this ability improves as embryos mature. Fort and his colleagues (p. 2323, see below) suggest that chemically disrupting the thyroid axis in turn disrupts retinoic acid effects on developing amphibian embryos. It has recently been shown that stress by itself alters concentrations of retinoic acid receptors in adult vertebrates (Brtko and colleagues, p. 233; see below).

Brtko, J., D. Macejova, J. Knopp and R. Kvetnansky. 2002. Effect of 2-deoxy-D-glucose—induced inhibition of glucose utilization on nuclear all-trans retinoic acid receptor status in rat liver. Pp. 233–235. *In* McCarty, R., G. Aguilera, E. Sabban and R. Kvetnansky (Eds.), Stress: Neural, Endocrine and Molecular Studies. Seventh Symposium on Catecholoamines and Other Neurotransmitters in Stress. Taylor and Francis, London.

Fort, D.J., R.L. Rogers, H.F. Copley, L.A. Bruning, E.L. Stover, J.C. Helgen and J.G. Burkhart. 1999. Progress toward identifying causes of maldevelopment induced in *Xenopus* by pond water and sediment extracts from Minnesota, USA. Environmental Toxicology and Chemistry 18:2316–2324.

Fulford, A.J. and D.S. Jessop. 2002. Neuropeptides in immune tissues: mediators of the stress response. Pp. 175–181. *In* McCarty, R., G. Aguilera, E. Sabban and R. Kvetnansky (Eds.), Stress: Neural, Endocrine and Molecular Studies. Seventh Symposium on Catecholoamines and Other Neurotransmitters in Stress. Taylor and Francis, London.

Goldstein, D.S., I. Elman, S.M. Frank and J.W.M. Lenders. 2002. Stressor-specific activation of catecholaminergic systems: clinical demonstrations. Pp. 249–253. *In* McCarty, R., G. Aguilera, E. Sabban and R. Kvetnansky (Eds.), Stress: Neural, Endocrine and Molecular Studies. Seventh Symposium on Catecholoamines and Other Neurotransmitters in Stress. Taylor and Francis, London.

Palkovits, M. 2002. Stress-related central neuronal regulatory circuits. Pp. 1–11. *In* McCarty, R., G. Aguilera, E. Sabban and R. Kvetnansky (Eds.), Stress: Neural, Endocrine and Molecular Studies. Seventh Symposium on Catecholoamines and Other Neurotransmitters in Stress. Taylor and Francis, London.

11. Souder, W. 2000. A Plague of Frogs: The Horrifying True Story. Hyperion Press, New York.

12. Faeth, P. and G.T. Mehan III. 2007. Nutrient runoff creates dead zone. World Resources Institute (accessed at http://pubs.wri.org/pubs_content_text.cfm?ContentID=3301).

13. Lannoo, M.J., D.R. Sutherland, P. Jones, D. Rosenberry, R.W. Klaver, D.M. Hoppe, P.T.J. Johnson, K.B. Lunde, C. Facemire and J.M. Kapfer. 2003. Multiple causes for the malformed frog phenomenon. Pp. 233–262. *In* Linder, G., S. Krest, D. Sparling and E. Little (Eds.), Multiple Stressor Effects in Relation to Declining Amphibian Populations. American Society for Testing Materials International, West Conshoshocken, Pennsylvania.

14. Blaustein, A.R. and P.T.J. Johnson. 2003. The complexity of deformed amphibians. Frontiers in Ecology and the Environment 1:87–94.

15. Blaustein, A.R. and P.T.J. Johnson. 2003. Explaining frog deformities. Scientific American, February:60–65.

16. Cohen, L. 1984. Hallelujah. Stranger Music, Incorporated, Sony Music, Canada.

SPECIES AFFECTED

1. Lin, M. 2000. Boundaries. Simon and Schuster, New York.

2. Lannoo, M.J., A.L. Gallant, P. Nanjappa, L. Blackburn and R. Hendricks. 2005. Introduction. Pp. 351–380. *In* Lannoo, M.J. (Ed.), Amphibian Declines: The Conservation Status of United States Species. University of California Press, Berkeley, California.

3. NARCAM (North American Reporting Center Amphibian Malformations) dataset at: www.frogweb.nbii.gov/narcam. Accessed on 16 October 2006.

4. Ouellet, M. 2000. Amphibian deformities: Current state of knowledge. Pp. 617–661. *In* Sparling, D.W., G. Linder and C.A. Bishop (Eds.), Ecotoxicology of Amphibians and Reptiles. Society for Environmental Toxicology and Contaminants (SETAC) Press, Pensacola, Florida.

5. Lannoo, M.J., D.R. Sutherland, P. Jones, D. Rosenberry, R.W. Klaver, D.M. Hoppe, P.T.J. Johnson, K.B. Lunde, C. Facemire and J.M. Kapfer. 2003. Multiple causes for the malformed frog phenomenon. Pp. 233–262. *In* Linder, G., S. Krest, D. Sparling and E. Little (Eds.), Multiple Stressor Effects in Relation to Declining Amphibian Populations. American Society for Testing Materials International, West Conshoshocken, Pennsylvania.

6. Ouellet, M., J. Bonin, J. Rodrigue, J.-L. DesGranges and S. Lair. 1997. Hindlimb deformities (ectromelia, ectrodactyly) in free-living anurans from agricultural habitats. Journal of Wildlife Diseases 33:95–104.

Rollo, C.D. 1995. Phenotypes: Their Epigenetics, Ecology and Evolution. Chapman and Hall, London, England.

7. Washburn, F.L. 1899. A peculiar toad. American Naturalist 33:139–141.

Crosswhite, E. and M. Wyman. 1920. [No title]. Journal of the Entomological Society 12:78.

Johnson, P.T.J., K.B. Lunde, R.W. Haight, J. Bowerman and A.R. Blaustein. 2001. *Ribeiroia ondatrae* (Trematoda: Digena) infection induces severe limb malformations in western toads (*Bufo boreas*). Canadian Journal of Zoology 79:370–379.

8. Lannoo, M.J. Personal collection from Hancock County, Indiana.

9. Heatwole, H. and P.J. Suárez-Lazú. 1965. Supernumerary legs in *Bufo marinus* and abnormal regeneration of the tail in *Ameiva exsul*. Journal of the Ohio Herpetological Society 5:30–31.

10. Listed at NARCAM site as *Bufo valliceps*, but see endnote 2 and Mulcahy, D.G. and J.R. Mendelson III. 2000. Phylogeography and speciation of the morphologically variable, widespread species *Bufo valliceps*, based on molecular evidence from mtDNA. Molecular Phylogenetics and Evolution 17:173–189.

11. Gray, R.H. 2000. Morphological abnormalities in Illinois cricket frogs, *Acris crepitans*, 1968–1971. Journal of the Iowa Academy of Science 107:92–95.

Blackburn, L.M. 2001. Status of Blanchard's cricket frogs (*Acris crepitans blanchardi*) along their decline front: Population parameters, malformation rates and disease. Master's thesis. Ball State University, Muncie, Indiana.

McCallum, M.L. and S.E. Trauth. 2003. A forty-three-year museum study of northern cricket frog (*Acris crepitans*) abnormalities in Arkansas: upward trends and distributions. Journal of Wildlife Diseases 39:522–528.

Smith, D.D. and R. Powell. 1983. *Acris crepitans blanchardi* (Blanchard's cricket frog): Anomalies. Herpetological Review 14:118–119.

12. Malformations previously unknown in this species; animals were collected as part of the U.S.F.W.S. National Wildlife Refuge survey.

13. Hebard, W.B. and R.B. Brunson. 1963. Hind limb anomalies of a western Montana population of the Pacific tree frog *Hyla regilla* Baird and Girard. Copeia 1963:570–572.

Johnson, P.T.J. and K.B. Lunde. 2005. Parasite infection and limb malformations: a growing problem in amphibian conservation. Pp. 124–138. *In* Lannoo, M.J. (Ed.), Amphibian Declines: The Conservation Status of United States Species. University of California Press, Berkeley, California.

Johnson, P.T.J., K.B. Lunde, E.G. Ritchie and A.E. Launer. 1999. The effect of trematode infection on amphibian limb development and survivorship. Science 284:802–804.

Johnson, P.T.J., K.B. Lunde, E.G. Ritchie, J.K. Reaser and A.E. Launer. 2001. Morphological abnormality patterns in a California amphibian community. Herpetologica 57:336–352.

Johnson, P.T.J., K.B. Lunde, E.M. Thurman, E.G. Ritchie, S.N. Wray, D.R. Sutherland, J.M. Kapfer, T.J. Frest, J. Bowerman and A.R. Blaustein. 2002. Parasite (*Ribeiroia ondatrae*) infection linked to amphibian malformations in the western United States. Ecological Monographs 72:151–168.

Miller, C.E. 1968. Frogs with five legs. Carolina Tips 31:1.

Reynolds, T.D. and T.D. Stephens. 1984. Multiple ectopic limbs in a wild population of *Hyla regilla*. Great Basin Naturalist 44:166–169.

Sessions, S.K. and S.B. Ruth. 1990. Explanations for naturally occurring supernumerary limbs in amphibians. Journal of Experimental Zoology 254:38–47.

Sessions, S.K., R.A. Franssen and V.L. Horner. 1999. Morphological clues from multilegged frogs: are retinoids to blame? Science 284:800–802.

See also: Johnson, P.T.J., J.M. Chase, K. L. Dosch, R.B. Hartson, J.A. Gross, D.J. Larson, D.R. Sutherland and S.R. Carpenter. 2007. Aquatic eutrophication promotes pathogenic infection in amphibians. Proceedings of the National Academy of Science 104:15781–15786.

14. Burkhart, J.G., J.C. Helgen, D.J. Fort, K. Gallagher, D. Bowers, T.L. Propst, M. Gernes, J. Magner, M.D. Shelby and G. Lucier. 1998. Induction of mortality and malformation in *Xenopus laevis* embryos by water sources associated with field frog deformities. Environmental Health Perspectives 106:841–848.

Hobson, B.M. 1958. Polymely in *Xenopus laevis*. Nature 181:862.

15. Banta, B.H. 1966. A six-legged anuran from California. Wasmann Journal of Biology 24:67–69.

16. Anonymous. 1944. Another abnormal frog. Turtox News 22:183.

Anonymous. 1954. Many-legged frogs. Turtox News 23:86–87.

Anonymous. 1962. Alabama biologist has 12-legged bullfrog. Bulletin of the Philadelphia Herpetological Society 10:24–25.

Houck, W.J. and C. Henderson. 1953. A multiple appendage anomaly in the tadpole of *Rana catesbeiana*. Herpetologica 9:76–77.

Lopez, T.J. and L.R. Maxson. 1990. *Rana catesbeiana* (bullfrog): Polymely. Herpetological Review 21:90.

Ouellet, M., J. Bonin, J. Rodrigue, J.-L. DesGranges and S. Lair. 1997. Hindlimb deformities (ectromelia, ectrodactyly) in free-living anurans from agricultural habitats. Journal of Wildlife Diseases 33:95–104.

Pearson, P.G. 1960. A description of a six-legged bullfrog, *Rana catesbeiana*. Copeia 1960:50–51.

Pelgen, J.L. 1951. A *Rana catesbeiana* with six functional legs. Herpetologica 7:138–139.

Rowe, C.L., O.M. Kinney, A.P. Fiori and J.D. Congdon. 1996. Oral deformities in tadpoles (*Rana catesbeiana*) associated with coal ash deposition: Effects on grazing ability and growth. Freshwater Biology 36:723–730.

Rowe, C.L., O.M. Kinney and J.D. Congdon. 1998. Oral deformities in tadpoles of the bullfrog (*Rana catesbeiana*) caused by conditions in a polluted habitat. Copeia 1998:244–246.

Ruth, F.S. 1961. Seven-legged bullfrog, *Rana catesbeiana*. Turtox News 39:232.

Volpe, E.P. 1970. Understanding Evolution. Second Edition. Brown, Dubuque, Iowa.

17. Anonymous. 1945. More many-legged frogs. Turtox News 23:86–87.

Bonin, J., M. Ouellet, J. Rodrique, J-L. Des Grandes, F. Gagné, T.F. Sharbel and L.A. Lowcock. 1997. Measuring the health of frogs in agricultural habitats subjected to pesticides. *In* Green, D.M. (Ed.), Amphibians in Decline: Canadian

Studies of a Global Problem. Herpetological Conservation, Volume 1, Society for the Study of Amphibians and Reptiles. St. Louis, Missouri.

Cooper, J.E. 1958. Some albino reptiles and polydactylous frogs. Herpetologica 14:54–56.

Duméril, A. 1865. Observations sur la monstruosité dite polymélie ou augmentation du nombre des membres chez des batraciens anoures. Nouvelles Archives du Museum d'Histoire Naturelle 1:309–319.

Duméril, A. 1865. Trois cas de polymélie (members surnuméraires) observes sur des batraciens du genre *Rana*. Les Comptes Rendus de l'Académie des Sciences 60:911–913.

Froom, B. 1982. Amphibians of Canada. McClelland and Stewart, Toronto, Ontario.

Lowcock, L.A., T.F. Sharbel, J. Bonin, M. Ouellet, J. Rodrigue and J.-L. Des Granges. 1997. Flow cytometric assay for in vivo genotoxic effects of pesticides in green frogs (*Rana clamitans*). Aquatic Toxicology 38:241–255.

Martof, B. 1956. Factors influencing size and composition of populations of *Rana clamitans*. American Midland Naturalist 56:224–245.

Ouellet, M., J. Bonin, J. Rodrigue, J.-L. DesGranges and S. Lair. 1997. Hindlimb deformities (ectromelia, ectrodactyly) in free-living anurans from agricultural habitats. Journal of Wildlife Diseases 33:95–104.

18. Cunningham, J.D. 1955. Notes on abnormal *Rana aurora draytoni*. Herpetologica 11:149.

19. Murphy, T.D. 1965. High incidence of two parasitic infestations and two morphological abnormalities in a population of the frog *Rana palustris* Le Conte. American Midland Naturalist 74:233–239.

Ryder, J.A. 1878. A monstrous frog. American Naturalist 12:751–752.

Tuckerman, F. 1886. Supernumerary leg in a male frog (*Rana palustris*). Journal of Anatomy and Physiology 20:516–519.

20. Adler, K.K. 1958. A five-legged *Rana* from Ohio. Ohio Herpetological Society Trimon Report 1:21.

Charles, H. 1944. Abnormal frog. Turtox News 22:179.

Colton, H.S. 1922. The anatomy of a five-legged frog. Anatomical Record 24:247–253.

Eigenmann, C.H. and U.O. Cox. 1901. Some cases of salutatory variation. American Naturalist 35:33–38.

Helgen, J., R.G. McKinnell and M.C. Gernes. 1998. Investigation of malformed northern leopard frogs in Minnesota. Pp. 288–297. *In* Lannoo, M.J. (Ed.), Status and Conservation of Midwestern Amphibians. University of Iowa Press, Iowa City, Iowa.

Johnson, R.H. 1901. Three polymelous frogs. American Naturalist 35:25–31.

Kingsley, J.S. 1881. A case of polymely in the Batrachia. Proceedings of the Boston Society of Natural History 21:169–175.

Merrell, D.J. 1969. Natural selection in a leopard frog population. Journal of the Minnesota Academy of Science 35:86–89.

Ouellet, M., J. Bonin, J. Rodrigue, J.-L. DesGranges and S. Lair. 1997. Hindlimb deformities (ectromelia, ectrodactyly) in free-living anurans from agricultural habitats. Journal of Wildlife Diseases 33:95–104.

Rosine, W.N. 1955. Polydactylism in a second species of Amphibia in Muskee Lake, Colorado. Copeia 1955:136.

Wagner, G. 1913. On a peculiar monstrosity in a frog. Biological Bulletin 25:313–317.

21. Anonymous. 1945. More many-legged frogs. Turtox News 23:86–87.

Gardiner, D.M. and D.M. Hoppe. 1999. Environmentally induced limb malformations in mink frogs (*Rana septentrionalis*). Journal of Experimental Zoology 284:207–216.

22. Lowcock, L.A., T.F. Sharbel, J. Bonin, M. Ouellet, J. Rodrigue and J.-L. Des Granges. 1997. Flow cytometric assay for in vivo genotoxic effects of pesticides in green frogs (*Rana clamitans*). Aquatic Toxicology 38:241–255.

23. Sessions, S.K. and S.B. Ruth. 1990. Explanations for naturally occurring supernumerary limbs in amphibians. Journal of Experimental Zoology 254:38–47.

24. Kingsley, J.S. 1880. An abnormal foot in *Amblystoma*. American Naturalist 14:594.

Winslow, G.M. 1904. Three cases of abnormality in urodeles. Tufts College Studies 1:387–410.

Worthington, R.D. 1974. High incidence of anomalies in a natural population of spotted salamanders, *Ambystoma maculatum*. Herpetologica 30:216–220.

25. Semlitsch, R.D., G.B. Moran and C.A. Shoemaker. 1981. *Ambystoma talpoideum* (mole salamander): morphology. Herpetological Review 12:69.

26. Bishop, D.W. 1947. Polydactyly in the tiger salamander. Journal of Heredity 38:290–293.

Bishop, D.W. and R. Hamilton. 1947. Polydactyly and limb duplication occurring naturally in the tiger salamander, *Ambystoma tigrinum*. Science 106:641–642.

Rosine, W.N. 1955. Polydactylism in a second species of Amphibia in Muskee Lake, Colorado. Copeia 1955:136.

Sealander, J.A. 1944. A teratological specimen of the tiger salamander. Copeia 1944:63.

In 1998, the Milwaukee Public Museum kept an adult five-legged tiger salamander. The duplication was at the right forelimb. The extra limb was normal size and projected dorsally. Interestingly, the flexor and extensor muscle groups of this limb were innervated identically to the normal limb on that side, and when the animal walked the abnormal limb extended and flexed—the animal appeared to wave.

27. Pendlebury, G.B. 1973. Distribution of the dusky salamander *Desmognathus fuscus fuscus* (Caudata: Plethodontidae) in Quebec, with special reference to a population from St. Hilaire. Canadian Field-Naturalist 87:131–136.

28. Lazell, J. 1995. *Plethodon albagula* (western slimy salamander): Foot anomalies. Herpetological Review 26:198.

29. Bonin, J., J.F. Desroches, M. Ouellet and A. Leduc. 1999. Les forêts anciennes: refuges pour les salamanders. Nature Canada 123:13–18.

Hanken, J. 1983. High incidence of limb skeletal variants in a peripheral population of the red-backed salamander, *Plethodon cinereus* (Amphibia: Plethodontidae), from Nova Scotia. Canadian Journal of Zoology 61:1925–1931.

Hanken, J. and C.E. Dinsmore. 1986. Geographic variation in the limb skeleton of the red-backed salamander, *Plethodon cinereus*. Journal of Herpetology 20:97–101.

30. Marvin, G.A. 1995. *Plethodon glutinosus* (slimy salamander): Morphology. Herpetological Review 26:30.

Winslow, G.M. 1904. Three cases of abnormality in urodeles. Tufts College Studies 1:387–410.

31. Marvin, G.A. and V.H. Hutchison. 1997. *Plethodon kentucki* (Cumberland Plateau woodland salamander): Morphology. Herpetological Review 28:199.

32. Dwyer, C.M. and J. Hanken. 1990. Limb skeletal variation in the Jemez Mountain salamander, *Plethodon neomexicanus*. Canadian Journal of Zoology 68:1281–1287.

33. Shubin, N., D.B. Wake and A.J. Crawford. 1995. Morphological variation in the limbs of *Taricha granulosa* (Caudata: Salamandridae): Evolutionary and phylogenetic implications. Evolution 49:874–884.

34. Hoppe, D.M. 2005. Malformed frogs in Minnesota: history and interspecific differences. Pp. 103–108. *In* Lannoo, M.J. (Ed.), Amphibian Declines: The Conservation Status of United States Species. University of California Press, Berkeley, California.

35. Johnson, P.T.J., K.B. Lunde, E.G. Ritchie, J.K. Reaser and A.E. Launer. 2001. Morphological abnormality patterns in a California amphibian community. Herpetologica 57:336–352.

36. Souder, W. 2000. A Plague of Frogs: The Horrifying True Story. Hyperion Press, New York.

INDEX